Peter Knief

CNT's & human lung epithelials in vitro, assessed by Raman Spectroscopy

Peter Knief

CNT's & human lung epithelials in vitro, assessed by Raman Spectroscopy

MatLab snippets in the hand of Raman Spectroscopy

Südwestdeutscher Verlag für Hochschulschriften

Imprint
Any brand names and product names mentioned in this book are subject to trademark, brand or patent protection and are trademarks or registered trademarks of their respective holders. The use of brand names, product names, common names, trade names, product descriptions etc. even without a particular marking in this work is in no way to be construed to mean that such names may be regarded as unrestricted in respect of trademark and brand protection legislation and could thus be used by anyone.

Publisher:
Südwestdeutscher Verlag für Hochschulschriften
is a trademark of
Dodo Books Indian Ocean Ltd., member of the OmniScriptum S.R.L Publishing group
str. A.Russo 15, of. 61, Chisinau-2068, Republic of Moldova Europe
Printed at: see last page
ISBN: 978-3-8381-2540-4

Zugl. / Approved by: Dublin, DIT, Diss., 2011

Copyright © Peter Knief
Copyright © 2011 Dodo Books Indian Ocean Ltd., member of the OmniScriptum S.R.L Publishing group

Table of Contents

Chapter 1 : Nanotechnology and Nanomaterials — 11

    1.1 Introduction — 11

    1.2 Carbon Nanotubes — 12

    1.3 Toxicology of Nanoparticles — 14

    1.4 How are we exposed? — 16

    1.5 Assessment of toxicity of CNTs to date — 17

    1.6 Why Spectroscopy? — 18

    1.7 Aim of the Study — 19

    1.8 Summary — 20

Chapter 2 : General Materials & Methods — 22

    2.1 Introduction — 22

    2.2 Raman Spectroscopy — 22

    2.3 Raman Instrumentation — 26

    2.4 Materials & Methods — 27

        2.4.1 Substrate — 27

        2.4.2 Cell line — 28

        2.4.3 Cell Culture — 29

        2.4.4 Nano-particles — 29

        2.4.5 Exposure — 30

        2.4.6 Experimental measurement setup — 30

        2.4.7 Data Analysis platform — 31

    2.5 Data Processing — 31

        2.5.1 Data Preprocessing — 31

        2.5.2 Univariate Data analysis — 57

2.5.3 Multivariate Data Analysis   61

2.6 Summary   70

Chapter 3 : Exposure Study   70

3.1 Introduction   70

3.2 Materials and Methods   71

    3.2.1 Cell culture   71

    3.2.2 Spectroscopy   71

    3.2.3 Data analysis   73

3.3 Results   75

    3.3.1 Univariate Analysis   75

    3.3.2 Multivariate Analysis   81

    3.3.3 Results from hyper spectral Imaging 87

3.4 Conclusions   89

Chapter 4 : Secondary toxic responses as a result of medium depletion   91

4.1 Introduction   91

4.2 Materials and Methods   92

    4.2.1 Sample preparation   92

    4.2.2 Spectroscopy   93

    4.2.3 Data pre processing   93

4.3 Results   94

    4.3.1 Influence on the medium   94

    4.3.2 Impact of medium depletion on the cells   97

4.4 Discussion and Conclusions   114

Chapter 5 : Study of toxic responses as a result of Reactive Oxygen Species — 115

    5.1 Introduction — 115

    5.2 Materials and Methods — 116

        5.2.1 Sample preparation — 116

        5.2.2 Spectroscopy — 117

        5.2.3 Data analysis — 117

    5.3 Results — 119

        5.3.1 Visual Changes — 119

        5.3.2 Spectroscopic Changes — 122

        5.3.3 Multivariate Analysis — 128

        5.3.4 Multivariate PLS regression modeling — 136

    5.4 Discussion and Conclusions — 140

Chapter 6 : Discussion and conclusion — 142

    6.1 Introduction — 142

    6.2 Nano-toxicity and Spectroscopy — 142

    6.3 Improved Methods — 143

    6.4 Implications of the direct exposure — 144

    6.5 Consequences of indirect exposure — 145

    6.6 Implications of hydrogen peroxide exposure — 147

    6.7 Conclusion and future aspects — 148

References — 152

Appendix I — 165

    Books — 165

    Articles — 165

Presentations / Workshops	167
2007	167
2008	167
2009	167
Appendix II	**168**
Articles in preparation	168
Fundamental Matlab code snippets	169
intensity calibration algorithm	169
linearization of spectra	170
dynamicss assisted moving average noise filtering	171
histogram assisted noise removal	173
rubberband baseline removal	175
automated simultaneous curve fitting with multiple mixed G-L curves	179

Table of Figures

Figure 1-1 Schematic illustration of Single & multi walled Carbon nanotubes, capped with halves of Buckminster fullerenes (http://mavimo.org/chimica/nanotubi)   3

Figure 1-2 Scheme of coordinates of a CNT, (T) describes the tube axis , a_1 and a_2 describe the unit vector of the planar graphene honeycomb. (http://en.wikipedia.org/wiki/Carbon_nanotube)   4

Figure 2-1 Schematic virtual energy level diagram of Rayleigh and Raman scattering (Stokes & anti Stokes) (http://en.wikipedia.org/wiki/Raman_scattering)   17

Figure 2-2 Raman spectrum (Stokes & anti-Stokes) of a silicon crystal (nonimmersed, exited with 514.5 nm for one second with ~0,6mW and 1800l/mm grating)   20

Figure 2-3 Image of Horiba Jobin Yvon SA Labram Raman Spectrometer HR 800 UV   23

Figure 2-4 Microscopic Image of CCL-185 A549 human alveolar, cancerous lung cells of the American type culture collection.   25

Figure 2-5 Sixth-order certified polynomial for SRM 2243. The x-axis is expressed in Raman shift (cm^{-1}) relative to 514.5 (19435.18 cm^{-1})   31

Figure 2-6 oscillation of abscissa-interval of a recorded signal   32

Figure 2-7 Zoomed spectrum of a quartz substrate background, original spectrum (blue), linearized spectrum (red)   32

Figure 2-8 Shift detection via cross correlation (shifted spectrum (red) +10, top), cross-correlation intensity maximum at 10, bottom.   33

Figure 2-9 Influence of S/G filter settings used in literature on a cellular spectrum of A549 in the region 970cm-1-1070 cm-1   38

Figure 2-10 Influence of RMS filter settings on a cellular spectrum of A549 in the region 970cm^{-1} - 1070 cm^{-1} 39

Figure 2-11 Comparison between SG, MA (1st order window 5 points) and RMS (window 5 points) filters on a cellular spectrum of A549 in the region 970cm^{-1} - 1070 cm^{-1} 40

Figure 2-12 DRMS Filter applied to an A549 spectrum between 870 cm^{-1} - 1070cm^{-1}, original (blue),1.run (green), 2.run (red), 3.run (cyan)   41

Figure 2-13 Histogram and binning result (binned features =23, outside of the histogram) of a baseline overlaid with a sine function representing noise   43

Figure 2-14 Histogram and binning result (binned features = 300) of a spectrum of A549 ($850 cm^{-1}$ -$1170 cm^{-1}$)   44

Figure 2-15 Baseline removal with the rubberband method. Original spectrum (top), baseline removed spectrum (bottom)   47

Figure 2-16 Substrate (B) removal by subtraction of a recorded substrate (C, SNR_{dB} 24.88) from a Cellular A549 spectrum(A, SNR_{dB} 21.73), and a recorded substrate scaled to the signal (D, SNR_{dB} 23.79)   50

Figure 2-17 Boundaries of the fit parameters above (red) and below (blue) the mean substrate signal (green)   53

Figure 2-18 Example of the removal of the water feature at ~1640 cm^{-1} from a cellular spectrum of an A549 cell   56

Figure 2-19 Schematic Illustration of PCA   61

Figure 2-20 Scattered dataset (left) and dendrogram of a hierarchical clustering with Euclidean distance as distance metric (right) (http://en.wikipedia.org/wiki/Cluster_analysis)   67

Figure 2-21 Comparison between PCA and ICA (Apo Hyvärinen, 2000)   69

Figure 3-1 Micrograph of A549 exposed to 100 mg/l SWCNT with clearly visible aggregates (indicated by the arrows, the squares indicate the manually selected measurement acquisition points)   74

Figure 3-2 Raman Spectrum of SWCNT with characteristic features (RBM's at ~180-300 cm^{-1}, D-Line at ~1350 cm^{-1}, G-line at~1590 cm^{-1})   76

Figure 3-3 Raman Spectra of A549 exposed Cells (25 mg/l SWCNT) (a) and A549 control cells (b), filtered with Savitzky Golay Filter order 3, 15 points.   78

Figure 3-4 Intensity of G-Line at $1598 cm^{-1}$ versus concentration (left), Intensity of Amide I at $1656 cm^{-1}$ versus concentration (right)   80

Figure 3-5 Peak ratios of 1287/Amide III (A), 1302/Amide III (B), 1338/Amide III (C),versus concentration and 1338/Amide III (D) versus the G-line intensity   81

Figure 3-6 Correlation of the $1287 cm^{-1}$/Amide III peak ratio with colony size endpoint.   82

Figure 3-7 Individual principal component loadings plot of the first 5 components, (PC_1-PC_5 with explained variance of 68.2%, 20.3%, 4.5%, 2.5%, 1.6%)   84

Figure 3-8 Principal component score plot of PC_2-PC_3 for every exposed (red) and control population (blue)   85

Figure 3-9 Principal Component Score Plot of PC_2-PC_4 for every spectrum of exposed concentration (control, 1.56, 6.25, 25.0, 100.0 mg/l) after being doubly derivatized 86

Figure 3-10 Cross validation results, the lowest RMSECV was observed at 10 latent variables (RMSECV = 2.53). 87

Figure 3-11 GA optimised PLS regression model correlating Raman spectra to clonogenic endpoints 89

Figure 3-12 Phase contrast micrograph (200x) of A549 cells following 24 h exposure to 800 µg/ml SWCNT with 5% serum showing aggregates (arrows) on cell surface. These aggregates were still observed on the cell surface after several washes with PBS [Davoren et al. 2006] 90

Figure 3-13 Spatial distribution map of CNT influence on A-549 as measured by Raman spectroscopy at indicated by the D-line feature (1331.79cm^{-1}) top row and the G-Line feature (1590.77 cm^{-1})bottom row 91

Figure 4-1 Photo of depleted media samples at a range of concentrations (notation in mg/L) 100

Figure 4-2 Raman spectra of filtered medium depleted by a series of SWCNT concentrations. 100

Figure 4-3 Reduced Amide Intensity with increased concentration of SWCNT (n=3 per concentration) 102

Figure 4-4 Mean spectra of A549 lung cells exposed to medium depleted by different concentrations of SWCNT (n > 33) 103

Figure 4-5 Peak ratios of 1287/Amide III (A), 1302/Amide III (B), 1338/Amide III (C), versus concentration 105

Figure 4-6 Variation of the CH_2 feature at ~1450 cm^{-1} versus depletion concentration as a measure for overall lipid and protein content (error bars indicate the SD per datapoint). 106

Figure 4-7 Extracts (I) of peak intensity ratios versus CH_2 (1450 cm^{-1}) with the filter control (0 mg/L SWCNT depletion) set to 100 %. 107

Figure 4-8 Extracts (II) of peak intensity ratios versus CH_2 (1450 cm^{-1}) with the filter control (0 mg/L SWCNT depletion) set to 100 % 109

Figure 4-9 Scatterplot of the mean spectra of the depleted medium exposed cells after PCA, with PC1 describing 69% of the total variance and PC2 describing 14% total variance of the dataset. 112

Figure 4-10 Loadings plot of the first two PC's of the depleted medium dataset (PC_1 describing 69% variance and PC_2 describing 14% variance) with shaded regions of interest. **114**

Figure 4-11 Scatterplot of the mean spectra of the control and filter control medium exposed cells after PCA, with PC_1 describing 66% of the total variance and PC_2 describing 18% total variance of the dataset. **117**

Figure 4-12 First PC describing 66% of variance of the control and filter control medium exposed cells after PCA. **117**

Figure 4-13 Illustration of suggested viability response in dependence of starvation time and intensity expressed in % of viability (96h time point highlighted). **119**

Figure 4-14 Cytotoxicity of Arc Discharge SWCNT filtered medium to A549 cells after 24, 48, 72 and 96 h exposures determined by the AB assay. Data are expressed as percent of control mean ± S.D. of three in dependent experiments. (*) Denotes significant difference from control ($p \leq 0.05$). (taken from Casey et al., 2008) **120**

Figure 5-1 Microscopic images of A549 cells exposed to 50µM H_2O_2 for different intervals (A=0h,B=1h,C=3h,D=6h). The overlaid lines indicate the positions of the spectral line maps measured from a=1 to b=11 **129**

Figure 5-2 Line map from (a) to (b) of an untreated, fixed A549 lung cell. The corresponding microscopic image indicates the direction and positioning of the line scan on the cell **131**

Figure 5-3 A549 cell exposed to ROS for 1h (line map) and the corresponding microscope image, illustrating the location of the spectral map. **132**

Figure 5-4 Average cellular spectra of ROS (induced by H_2O_2) exposed A549 lung cells (control, 1h, 3h, 6h) **133**

Figure 5-5 Integrated signal intensity response of A549 in time dependence to the exposure to ROS **135**

Figure 5-6 Intensity of the fitted Phenylalanine response of A549 in time dependence to the exposure to ROS **136**

Figure 5-7 CH_2 feature at ~1449 cm^{-1} (fitted) in dependence of ROS exposure time. **136**

Figure 5-8 Variation of Ester bond at ~1740 cm-1 as a function of ROS exposure time **138**

Figure 5-9 Variation in time of toxicity markers (from Perna et. al.) of A549 lung cells exposed to 50 µM ROS (0-6 h) **139**

Figure 5-10 Scatterplot of scorings of the first 3 PC's 140
Figure 5-11 Scatterplot of scorings of the first 3 PC's rotated around the y-axis 142
Figure 5-12 Principal component loading plot of PC_1, PC_2 and PC_3 of ROS exposed A549 143
Figure 5-13 Ramam spectrum of dried phosphatidylcholine recorded with 514.5nm excitation for 10s with a grating of 1800 lines / cm 144
Figure 5-14 ICA scatterplot after whitening of scores on IC_1 - IC_3 145
Figure 5-15 Independent component loading plot of IC_1, IC_2 and IC_3 of ROS exposed A549 147
Figure 5-16 Comparison between PC_3 and IC_3 148
Figure 5-17 Micrograph of filter control medium exposed cells after 96 h exposure showing clear vesicularisation. 151
Figure 5-18 PLS regression prediction result for exposure time to ROS (full dataset) 152
Figure 5-19 PLS regression prediction result for exposure time to ROS (corrected dataset) 153
Figure 5-20 PLS regression - prediction result for predicted membrane protein damage to ROS (corrected & regrouped dataset) 155

Table of Tables

Table 2-1 Coefficients of the certified polynomial for 514.5 nm intensity calibration 32
Table 2-2 Noise Reduction Performance of common and proprietary noise filters 45
Table 2-3 List of the selected peaks to fit the substrate with their fit parameters. 55
Table 2-4 common peak assignments [91, 105, 147] 60
Table 2-5 Genetic Algorithm parameters 66
Table 3-1 Sample numbers after recording the measurements and outlier removal 75
Table 3-2 Performance of GA optimised PLS regression. 86
Table 4-1 Sample numbers (depletion study) after recording the measurements and outlier removal 94
Table 5-1 Sample numbers (ROS study) after recording the measurements and outlier removal 118

Chapter 1 : Nanotechnology and Nanomaterials

1.1 Introduction

Nanomaterials, structured components with at least one dimension less than 100 nm (US National Institute for Occupational Safety and Health) [1], are considered to be a new class of materials with unusual characteristics, not only due to the chemistry of the materials themselves, but because their dimensions have a significant impact on their chemical properties [2]. In addition, chemical modifications or functionalisation can change the optical, magnetic or electric properties of these materials [3]. Thin films and surface coatings used as features on computer chips or self cleaning glass are examples of nano materials with only one dimension in the nano scale. Examples of nano materials which have nano-scale extent in two dimensions are nanotubes and nano wires, while quantum dots and fullerenes are nano scale in three dimensions. Commonly, nanomaterials are produced and manipulated in one of two ways, either "top-down", starting from a bulk material mechanically or chemically downsized to a certain shape, or bottom-up, aggregated from smaller subunits to larger structures. Nano materials typically exhibit physical characteristics which are significantly different to their larger scale counterparts, due to the changed ratio of surface to volume and the increasing contribution of quantum effects [4]. Their chemical activity increases and renders them suitable for example as catalysts and optical coatings, as probes for the bio detection of chemicals, and as magnet resonance imaging (MRI) contrast agents [5]. All of these applications arise as the presence of quantum effects enhances the chemical, optical, magnetic or electrical properties of these materials. Nanomaterials have already seen an extraordinary range of applications in industry to date and are likely to find wide ranging applications in nano- science, -medicine and -engineering. Currently, nano-particles are used as composites and additives to cosmetics, clays and other convenience materials [6]. For example, titanium oxide nano-particles are used in sunscreens and in self-cleaning surfaces [7]. It is predicted that they will soon be commonly used in paints, fuel cells, additives, and other long-term applications as, for example, lubricants, medical implants and as nano engineered membranes [8]. In particular, zero-dimensional nano-particles such as Buckminster fullerenes or quantum dots are starting to play a key role in future medicine and science [9, 10]. Given potential or current applications, in drug delivery systems, reinforcements or coatings in implants or prosthetics, in chemical and molecular imaging [11], nanomaterials promise to become an essential part in the area of bio nano-technology and nano medicine.

1.2 Carbon Nanotubes

Carbon nanotubes, in both their multi- and single-walled forms, have attracted significant attention since their first emergence in 1991, having been first observed by Sumio Iijima [12]. They are one dimensional macro molecules of rolled graphene sheets, differentiated into single or multi walled forms, with diameters of the order of nanometers, and a length up to several centimetres. Single walled carbon nanotubes (SWCNTs) basically consist of a singular graphene cylindrical wall with a diameter of 0.8 to 1.6 nm [13] (Figure 1-1) and lengths up to four centimetres [14], whereas multi walled carbon nanotubes have multiple cylindrical walls of graphene in a coaxial alignment. The diameter of these multi walled carbon nanotubes is dependent on the number of graphene layers (Figure 1-1). Simplistically, they can be described as elongated Buckminster Fullerenes.

Figure 1-1 Schematic illustration of Single & multi walled Carbon nanotubes, capped with halves of Buckminster fullerenes (http://mavimo.org/chimica/nanotubi)

The way the graphene sheet is wrapped is described by the chiral vector (C_h), determined by a pair of indices (n, m). The scalars n and m denote the number of unit vectors along two directions in the honeycomb crystal lattice of the graphene sheet. If m equals 0, the nanotubes are called "zigzag" because of the circumferential zigzag structure. Similarly, if n equals m, the nanotubes are called "armchair". Otherwise they are called "chiral" (Figure 1-2). Due to the symmetry and the unique electronic structure of graphene, the structure of a nanotube strongly affects its electrical properties. For a nanotube of given (n, m), if n - m is a multiple of 3, then the nanotube is metallic, otherwise the nanotube is a semiconductor. Therefore all "armchair" (n=m) nanotubes are metallic, and otherwise, the nanotubes (5,0), (6,4), (9,1), etc. are semiconducting.

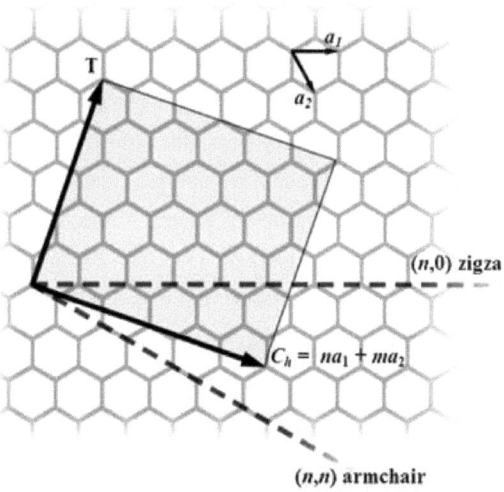

Figure 1-2 Scheme of coordinates of a CNT, (T) describes the tube axis, a_1 and a_2 describe the unit vector of the planar graphene honeycomb. (http://en.wikipedia.org/wiki/Carbon_nanotube)

Theoretically, metallic nanotubes can have an electrical current density more than 1.000 times greater than conventional metals such as silver and copper [15]. Various methods for the synthesis of carbon nanotubes have been developed. These methods range from electric arc discharge, laser ablation, and catalytic decomposition of hydrocarbons to high pressure carbon oxide conversion [12, 16-18]. Every method produces nanotubes of various dimensions and physical properties [19, 20]. As produced, a given volume of carbon nanotubes usually contains production specific impurities (e.g. traces of the catalytic materials required for production), and the SWCNTs are aggregated as bundles. At present, purification processes are not fully effective, adding further variation to the content of nanotube samples. Even after purification, the trace level of impurities cannot be ignored [21]. The separation of single tubes from these bundles, also after optimal purification, is problematic [22-24].

Nevertheless, because of their outstanding properties, a huge number of novel applications for carbon nanotubes is envisaged. Nanotubes are extremely light weight and very strong mechanically. Their elastic modulus is over 1 TPa [25], giving them the stiffness of diamond [26, 27]. They have excellent physical, thermal and chemical stability and conduct electricity very well [26]. According to the helicity of the rolled graphene sheet, carbon nanotubes display semiconductor or metallic properties and thereby offer a diverse range of applications in electronics and aviation or aerospace technologies [8], either as part of a molecular switch [28] or as a unit in quantum computers [29]. Due to the electrical conductivity of carbon nanotubes, electrically conducting polymers and tissue engineering constructs become feasible [30] Carbon nanotubes have opened other major research areas in the field of biomedical devices, including as the basis of nano-particle-based molecular and chemical imaging techniques [31, 32] and as nano sensors [33]. Single walled carbon nanotubes conjugated with single strand DNA fragments can possibly detect the presence of complementary DNA sequences by changing the conductivity of small sequences [34]. Nanotubes have also been used as the basis for the invention of nano devices facilitating biological functions. They are employed as repair mechanisms of macromolecules such as DNA, or as scaffolding materials for osteoblast proliferation to potentially replace lost biological functions [35]. Nanometer scaled carbon fibre composites have been shown to alter the selective adhesion of osteoblasts and change nerve cell functions and astrocyte adhesion [36]. They can be exploited as molecule transporters delivering agents directly into cells [37]. In order to improve the mechanical properties of existing materials, single walled carbon nanotubes can be used as nano sized filler compounds creating anisotropic nano composites [38]. As a result of all these potential applications, they are anticipated to see mass production in the near future, which naturally raises concern about their potential human and environmental impacts [39].

1.3 Toxicology of Nanoparticles

The same potential advantages of nano materials in terms of their relative size compared to single cells gives rise to fears of their potential harmful interaction with biological systems [40]. Although particle size and structure as a source of toxicity is not completely new (cf. asbestos), the EU takes an incremental approach in adapting the relevant regulations [41]. No mechanism is currently in place to limit the use of nanomaterials [42]. Ideally, however, before including these materials into new or existing devices and technologies, the toxicity and biocompatibility needs to be investigated [2]. Potential occupational hazards have to be investigated along with implementing the use of nanomaterials on a large scale [42]. Thus the field of 'nano-toxicology' can be defined as "The science of engineered nano devices and nano structures that deals with their effects on living organisms" [1]. Due to their increased reactivity and their nearly unlimited mobility, nanoparticles can be toxic even if the main constituent material (e. g. carbon or graphite in the case of carbon nanotubes) is considered harmless [1]. One of the dominating effects in the toxicity of nanoparticles is the induced generation of reactive oxygen species (ROS) [43], not only extracellularly, but also intracellularly. ROS is a global term given to a variety of species that are either neutral radicals, radical ions, or bound radicals of oxygen, nitrogen and other species. Neutral radicals (e.g. hydrogen peroxide) do not necessarily react by the radical-mediated pathways but have the potential to generate free radicals through Fenton reactions [44] and are therefore commonly employed to induce oxidative damage. Despite the fact that ROS are a chemical hazard, they play a major role in cell signalling and cell metabolism [45, 46]. Therefore, in cellular metabolism the handling of ROS is a well balanced process [47]. As soon as this equilibrium is changed the cellular metabolism is under oxidative stress, which can result in short and long term damage to exposed cells [48]. The possible damage mechanisms range from lipid peroxidation, conformational changes in proteins, enzymes and membranes of cellular organelles to nuclear factor activation, gene transcription and protein expression [45, 49, 50] and finally in the longer term, to pro inflammatory cytokine release [51] and mediated cell death.

Secondly, it is not necessarily the nano-particle itself which has a toxic influence on the organism. Secondary effects or indirect toxicity can play a major role in the overall toxic impact [52]. Nano-particles have several demonstrated secondary effects on organisms, which potentially block certain pathways or intervene in metabolic cycles [53, 54]. They may cause the overloading of the capacity of the organisms to tolerate their presence [55]. *In vitro*, they can interfere with the environment and limit the availability of nutrients by adsorbing them [52, 56]. The latter effect has been shown to virtually deplete the cell culture medium *in vitro* and thereby starve the exposed cells [52]. This starvation, a secondary effect, or termed here as indirect toxicity, finally could lead to a reduced viability or reduced proliferative capacity [57].

Therefore, the consequences of primary and secondary toxicity of the exposure to nano particles can be manifold. This complexity increases considering the impurities contained in the produced nano particles as another source for nano toxicity in terms of occupational exposure, and the possible interaction of primary and secondary effects between nanoparticles and their impurities.

1.4 How are we exposed?

Depending on the method of production or their application, nanoparticles can enter the body either by inhalation, ingestion or even by penetration of the skin. In therapeutic applications, nanoparticles might be directly injected into the body or presented to the digestive system [1]. It seems unlikely that humans will be exposed to manufactured nano-particles in doses detrimental to health, for example by pollution, but they might be inhaled in significant doses in certain environments [58], during production or during exposure to particles released from wear of coatings etc. [2, 40, 59]. The human lung, an organ that exchanges 10-20 m^3 of air on average per day [60], is likely to receive a much larger dose than any other organ. Therefore the lungs represent the primary channel of exposure for nano-particle dusts, resulting in their deposition in the airway on alveolar epithelium, mucus or epithelial lining fluid [1, 61]. These nano-particles on the one hand may be cleared by macrophages or by mucus exchange, or on the other hand could enter the intracellular space coming in contact with fibroblasts, endothelial cells or cells of the immune system [42, 55]. The skin, as the largest human organ, represents a further major route for both occupational and environmental exposure, in particular during the process of production of nanoparticles. Occasionally, particles that lodge within the avascular epidermis may not be removed by phagocytosis and could penetrate the stratum corneum [1]. Finally, through medical investigations it has been proven that some nanoparticles are phagocytosed in the human lung, liver and spleen, depending on their solubility and functionalisation, whereby they then may be excreted through the kidneys [61-63].

1.5 Assessment of toxicity of CNTs to date

The dominant reason for the toxicity of nano-particles is based on their physico-chemical properties. The chemical activity of a nanoparticle is not only determined by the bulk material itself. The dimensions and thus the relative surface area also play an important role in their chemical reactivity [64]. An increased surface area results in greater cellular interfacing opportunities and a larger bonding area for possible toxic interactions [1]. Nanoparticles have also been observed to increase reactive oxygen species (ROS) production themselves [65]. The physical dimensions and properties of the nano-particle themselves dictate the limitation of uptake in cells and organs and the possibility of phagocytosis [66]. Depending on the size of the nano-particle, they might pass the natural defences of the body, or might be just too big to be phagocytosed [55]. The retention time in the organism is dependent on their solubility and polarity, and therefore the longer the particle interacts with cellular membranes, the more intense are the toxic reactions and associated effects [67-69]. In studies to date, a whole battery of toxicological assays has been employed in order to assess the toxicity and toxicological endpoints of biological materials exposed to carbon nanotubes. As particle inhalation is considered to be one of the highest risk potential exposure scenarios, many studies have focussed on simulating lung exposure. The epithelial surface area of the lung is predominantly made up of alveolar type I cells. Lacking a reliable human *in vitro* model for this cell type, alveolar epithelial type II cell lines such as the A549 cell line are often used as a substitute for *in vitro* models [70]. Individual aspects of SWCNT exposure, like the reactive oxygen species production by carbonaceous aggregates [71, 72], translocation of nano-particles [73] and immune-reactivity [74] have already been studied on this cell line alone. In exposures to minimally processed SWCNTs, simulating occupational exposure to dust, no cellular internalization of the nanotubes was observed, indicating that the SWNT samples do not elicit a primary toxic response [75]. It has furthermore been shown that during cell viability studies of A549 human alveolar carcinoma cells, the carbon nanotubes interact with commonly used indicator dyes like, Coomassie Blue, Alamar Blue™, Neutral Red, MTT and WST-1 [76]. Spectroscopic analysis of the nanotube interactions demonstrated conclusively the interaction with the colorimetric dyes, compromising the associated optical activity commonly used to evaluate particle toxicity [76]. A549 cells have also been employed for clonogenic assays, to assess viability post exposure to nano particles [77]. The clonogenic assay has proven to be a valid alternative to assess cell viability and proliferative capacity [78], assessing the toxic response by monitoring colony number and size, thereby giving an overall response to exposure [79]. The application of the clonogenic assay in viability studies of epithelial lung cells exposed to single wall carbon nanotubes reveals significant results. The clonogenic endpoints of colony size and colony number have demonstrated an overall toxic response as a result of exposure to carbon-based materials. It is notable that the colony number, normally an endpoint associated with viability, was not significantly influenced, whereas the colony size, an indication of proliferative capacity, was significantly reduced compared to controls. The influence on the proliferative capacity has been associated with a secondary effect due to the depletion of the medium as a result of the adsorption of medium components onto the surface of the carbon-based nano materials [80]. This adsorptive nature of the nanotubes and the reduced colony size suggest that the cells were more likely starved due to depletion of the nutrient medium. In terms of toxicity, this reveals a secondary effect, already known from nutrient deficient environments [77, 81]. Although this technique provides accurate results, the assay is very time and resource-consuming in terms of setup and analysis and provides only information about the growth of cell cultures [82] rather than more specific intracellular mechanisms of response.

1.6 Why Spectroscopy?

Raman spectroscopy is explored in this project as a potential toxicity screening method to overcome the difficulties experienced in the application of classical toxicological assays for the assessment of nanoparticles [76]. Ultimately, it can potentially revolutionise the screening techniques, providing details of cellular response at a molecular level, with subcellular resolution. It is a cost effective method that requires little effort in sample preparation and no additional assay to measure an effect. Raman spectroscopy is a very versatile analytical tool, known for its strengths in the physical and chemical characterization of materials and systems [83-85]. The modality potentially offers analytical and diagnostic information at a high sub cellular spatial resolution [86]. It derives an additional benefit from the minimal need for processing of biological materials. For example, one can perform Raman measurements in solution, whether that solution is opaque or translucent, or contains strongly reflecting or strong absorbing materials [87-89]. It has already been shown to be a viable tool for disease diagnosis [90] as well as for the detection of extracellularly mediated changes in the chemical content of the cell [91]. The Raman spectrum of a cell also contains chemical information regarding its constituents, providing a complete biochemical fingerprint of the cell, and ultimately exhibiting signatures that may be indicative of cell state, e.g. proliferation, apoptosis, necrosis, etc. [90, 92-94]. Though changes to single spectral features, as a result of any changes to the state of the cell or a molecule, are possible, it is more often a complex effect that is exhibited by a number of dependent and independent spectral features. Since the molecular content of the cell is so complex, the Raman spectra of cells and tissues are complex convolutions of the vibrational signatures of each of the components. Therefore, multivariate analysis (MVA) of spectroscopic data is required to analyze the complex spectral changes resulting from changes in the state of the cell after exposure to external agents. MVA delivers a detailed view of the overall response [95, 96] and allows for example the classification of pathologically altered tissues. Multivariate analysis can quantify the response of biological materials to external stimuli. Depending on the method used, it is possible to model the spectral features for characterization of biological effects [97, 98].

1.7 Aim of the Study

The aim of this study is the development of Raman spectroscopy with advanced signal processing technologies and to explore its application as a tool to assess the toxicity responses, or more generally the biological responses, in cells exposed to potentially toxic nanomaterials. In particular, the study aims to develop and elucidate methodologies for the analysis of the toxic effects of carbon nanotubes on human lung cells *in vitro*. The study specifically duplicates previous toxicological assessments with the aim of correlating spectroscopic signatures with established gold standard techniques. The SWCNTs samples are minimally processed, with the objective of mimicking the real occupational exposure of human epithelial lung cells to airborne carbon nanotube dusts. The goal is to comprehensively characterize the possible toxic effects by evaluating aspects of the primary or secondary toxicity with Raman spectroscopy. It is expected that this technique can overcome the existing problems of common toxicological assays in terms of informative value, rapidity, ease of sample preparation, and thus cost of application. In the initial study presented in chapter 3, A549 cells were exposed to various doses of minimally processed SWCNT's, according to the exact protocol previously employed for clonogenic studies [99]. The results are compared and correlated with the results of this clonogenic study using both uni and multi variate analyses in order to establish the robustness of the technique. Although clonogenic endpoints are employed as the gold standard of toxicological assays [78, 82, 100] they are non-specific, and consequently in chapter 4 the study is extended to explicitly monitor secondary effects due to medium depletion, caused by the adhesion of the nutrients to the SWCNTs. As already demonstrated by Casey et al. [101], employing colorimetric and clonogenic methods, carbonaceous nano particles reveal significant cytotoxicity, due to alteration of cell culture medium, at high exposure doses only.

Induced immuno responses have also been observed as a result of exposure the SWCNTs and are characterised by the release of cytokines. In inflammatory responses of A549 lung cells, ROS are known to be an aggressive mediator. Single wall carbon nanotubes on the other hand, have been demonstrated to be a significant source of oxidative stress [1] and therefore primary toxicity.. Therefore, in an attempt to determine the spectroscopic signature of this primary toxicity and the influence of ROS on the cell viability, in chapter 5, A549 cells are exposed directly to medium enriched with reactive oxygen species as a result of exposure to hydrogen peroxide.

The results of all spectroscopic studies are compared directly with conventional assays published in the literature [45, 52, 76, 102-104]. Spectroscopic signatures of a range of specific toxicological responses are ascertained. As a bonus, it is expected that, with extensive two-dimensional spatial recordings of exposed cells, the localization of carbon nanotubes and their toxicological impact can be elucidated.

1.8 Summary

Nano-particles are already common constituents of modern technology. They may be used not only in technical but also in medical applications in the future. Their reactivity is disproportionally higher than bulk carbon and it has already been shown that they can have a huge biological impact. Thus, it is vitally important to analyze the toxicological aspects of exposure to nano-particles. The targeted tissue and therefore the *in vitro* cell line should match the route of exposure. The means of analysis should be influenced as little as possible by the nano-particle itself, however. The common cyto-toxicological assays have been shown to be problematic in terms of assessing the results of exposure to carbon nanotubes whereas the clonogenic assay seems to be a more reliable standard to compare to the expected results. It is, however, non specific in terms of mechanisms of interaction and Raman spectroscopy is proposed as an alternative. Raman spectroscopy is established as a versatile tool in materials sciences. This technique is now being adapted to the analysis of the effects of exposure in biological materials. The advantage is the minimal requirement for sample preparation and the rapidity of measurement combined with high spectral and spatial resolution. It enables more realistic experimental conditions to mimic the real exposure of human epithelial lung tissue to carbon nanotube dust. Within this thesis, studies demonstrating a correlation of the non-specific toxicological response with conventional cyto-toxicological methods are described. The overall aim of the study is to establish the spectroscopic responses associated with more specific responses.

Chapter 2 : General Materials & Methods

2.1 Introduction

In this chapter, the general spectroscopic, sample preparation, data pre processing and data analysis techniques employed in this work are outlined. The physics of Raman spectroscopy and its chemical and physical applications are elucidated. The influences of the substrate employed, the exposed cell line, and the test materials on the experiment will be discussed. The methods of exposure, preparation of the samples, storage and the actual measurements are described. Parallel to considerations of linearity, noise, and baseline, aspects of the derived spectra and corresponding pre processing methods, the data processing techniques used are outlined. Mathematical data processing is introduced to cover univariate and multivariate methods. Finally, chemical imaging in Raman spectroscopy based on the previously described techniques is introduced and a sample of the current studies is given.

2.2 Raman Spectroscopy

Upon interaction with a material, light can be absorbed, reflected or scattered. Rayleigh scattering (elastic scattering) occurs when the scattered light is of the same frequency as the incident light. Raman scattering (inelastic scattering) is a result of light that is scattered by a material (molecule in the case of the current study) whereby its frequency differs from that of the incident light as a result of the interaction of the photon with the molecular vibrations (Figure 2-1). In Raman scattering, the energy increase or decrease from the excitation is related to the vibrational energy spacing in the ground electronic state of the molecule and therefore the Raman shift of the Stokes and anti-Stokes lines are a direct measure of the vibrational energies of the molecule. In Stokes Raman scattering, the molecule starts out in a lower vibrational energy state and after the scattering process ends up in a higher vibrational energy state. Thus, the interaction of the incident light with the molecule creates a vibration in the material, and the scattered photon is reduced in energy.

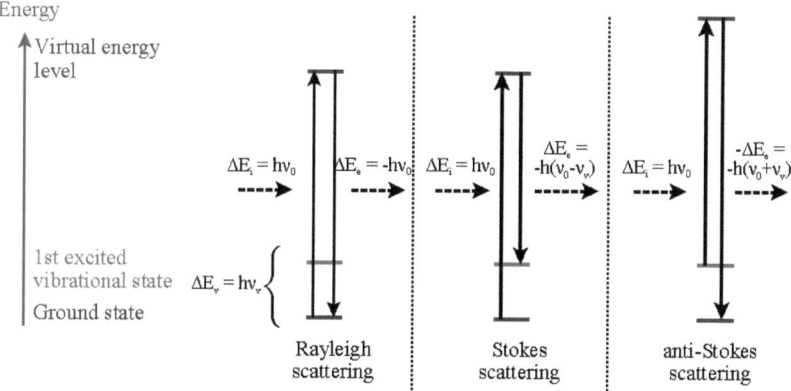

Figure 2-1 Schematic virtual energy level diagram of Rayleigh and Raman scattering (Stokes & anti Stokes) (http://en.wikipedia.org/wiki/Raman_scattering)

In anti-Stokes scattering, the molecule begins in a higher vibrational energy state and after the scattering process ends up in a lower vibrational energy state. Thus a vibration in the material is annihilated as a result of the interaction and the scattered photon has an increased energy. The frequency differences between the Raman lines and the exciting line are characteristic of the scattering substance and are independent of the frequency of excitation. The Raman-effect arises from the coupling of the induced polarisation of scattering molecules which is caused by the interaction of the electric field vector of the electromagnetic radiation with the molecular vibrational modes. Light of frequency ω_L produces a polarization in a material given by Equation 2-1.

$$P(\omega_L) = \chi(\omega_L) E_0 \cos\omega_L t \qquad \text{Equation 2-1}$$

where P is the polarisation, ω_L is the frequency of incident light, E is the electric field and $\chi(\omega_L)$ is the polarizability or susceptibility, normally considered a constant of the material associated with its electronic properties. However, at a finite temperature, a material is not at equilibrium and atoms will vibrate about their equilibrium positions, R, along the normal coordinates, with frequency ω_K in accordance with a simple harmonic oscillator approximation. The displacement from equilibrium can be represented by Equation 2-2.

$$\Delta R(t) = \Delta R \cos(\omega_K t) \qquad \text{Equation 2-2}$$

The susceptibility to polarisation thus oscillates about its equilibrium value χ_0, and can be represented by Equation 2-3;

$$\chi_k(t) = \chi_0 + \Delta\chi_k \cos(\omega_k t)$$ Equation 2-3

The polarisation now has the form as illustrated in Equation 2-4;

$$P(\omega_L, \omega_K) = \chi_0(\omega_L)E_0 \cos\omega_L t + \Delta\chi_K E_0 \cos(\omega_L)t \cos(\omega_K - \delta_K)$$ Equation 2-4

where δ_K takes into account any phase difference between the molecular vibration and the electric field oscillation. This may be written as Equation 2-5:

$$P(\omega_L, \omega_K) = \chi_0(\omega_L)E_0 \cos\omega_L t + 1/2\Delta\chi_K E_0 (\cos((\omega_L - \omega_K)t - \delta_K) + \cos((\omega_L + \omega_K)t + \delta_K))$$ Equation 2-5

Thus the polarization has the form:

$$P = P(\omega_0) + P(\omega_0 - \omega_K) + P(\omega_0 + \omega_K)$$ Equation 2-6

An oscillating polarization will reradiate at the oscillation frequency and thus the scattered light has three components. $P(\omega_0)$ gives rise to Raleigh scattering. $P(\omega_0 - \omega_K)$ corresponds to the subtraction of a vibrational quantum from the photon energy and the creation of a vibration and gives rise to the "Stokes lines" of a Raman spectrum. $P(\omega_0 + \omega_K)$ corresponds to the addition of a vibrational quantum to the photon by the annihilation of a vibration and results in the "anti-Stokes lines" of a Raman spectrum.

$$\frac{I_S}{I_{AS}} = \left(\frac{\omega_0 + \omega_K}{\omega_0 - \omega_K}\right)^4 e^{\left(\frac{hv}{kT}\right)}$$ Equation 2-7

In Raman spectroscopy, each vibration can couple to the laser generating a vibrational spectrum on both Stokes and anti-Stokes sides. The Stokes shift is normally measured, as at room temperatures it is easier to create a phonon than to annihilate one, in accordance with the Boltzmann population distribution. The ratio between the Stokes and anti Stokes scattered intensities, as represented in Equation 2-7, can be used as a measure of the temperature of a material under irradiation. In Raman spectroscopy, the parameter of interest is the frequency shift as a result of the coupling to the vibrations and thus the incident frequency is set to zero and the Stokes line shift is represented as a positive shift. For comparison with infrared spectroscopy, frequency shifts are usually given in wavenumbers. Figure 2-2 shows for example the Raman spectrum of a Silicon crystal. Its sharp peaks are often used for calibration purposes in Raman spectroscopy. The Stokes (positive) and anti-

Stokes (negative) Raman lines can be seen symmetrically shifted from the incident laser line by a frequency of 520.7cm^{-1}.

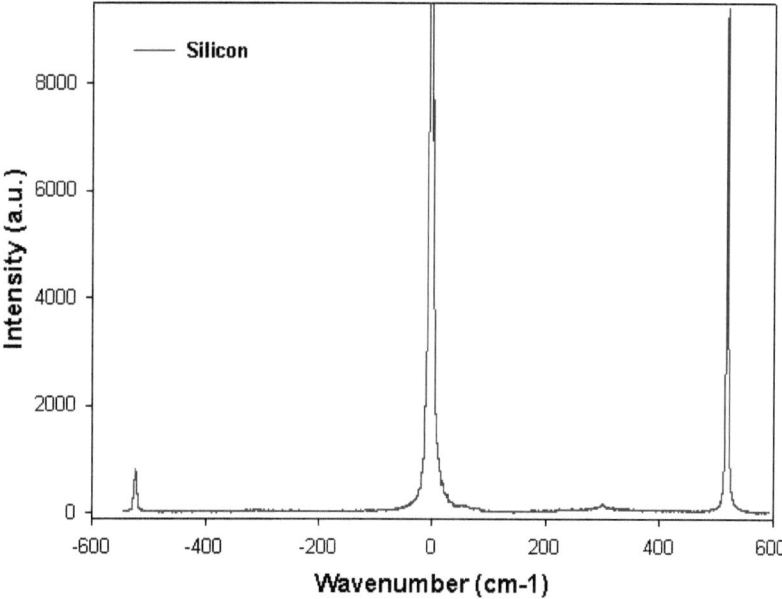

Figure 2-2 Raman spectrum (Stokes & anti-Stokes) of a silicon crystal (nonimmersed, exited with 514.5 nm for one second with ~0,6mW and 1800l/mm grating)

The selection rule for Raman activity dictates that there must be a change in the polarizability of a molecule during the vibration. For example, the symmetric stretch of CO_2, although IR inactive, is Raman active and the peak occurs at 1351cm^{-1} in its Raman spectrum. Therefore, although Raman spectroscopy is very similar to the more frequently used IR spectroscopic technique, the two vibrational spectroscopic techniques are complementary. For polar molecules, some vibrations give rise to strong bands in the IR spectrum since they cause large changes in the dipole moment, but they give weak Raman signals since it is difficult to induce a change in the polarizability of such molecules. Like-wise, for non-polar molecules, vibrations that give very strong Raman bands usually result in weak infrared signals. For example, hydroxyl or amine stretching vibrations, or the vibrations of carbonyl groups, are usually very strong in an IR spectrum, and usually weak in a Raman spectrum [105]. However, the stretching vibrations of carbon double or triple bonds and symmetric vibrations of aromatic groups are very strong in the Raman spectrum.

Raman spectroscopy has the advantage of minimal interference from the polar OH vibrations of water so it is a good choice for biological samples with a view towards *in vivo* measurements. Also, as Raman is usually carried out using visible wavelengths, higher spatial and spectral resolutions can be achieved. Currently vibrational spectroscopy utilizing Raman spectroscopy for the characterization of biological specimens has a large number of demonstrated applications. Since the year 2000, more than 1980 articles dealing with Raman spectroscopy for the purposes of characterization of biological specimens have been listed in Pub MED Central (PMC) alone and the number increases annually. A large number of *in vitro* and *in vivo* applications have been developed to date [86, 94, 106]. The development of the modality as a means of assessing the cyto-toxicological impact in biological specimens is therefore a natural progression [95, 107, 108].

2.3 Raman Instrumentation

The Instruments SA Labram 1B and Horiba Jobin Yvon Labram HR800 UV utilised in this work are confocal Raman imaging microscope systems. Both have motorised XY sample stages for automated Raman imaging. The Labram1B has both Helium-Neon (632.8nm/11mW) and Argon ion (514.5nm/130mW), lasers available as excitation sources. The Labram HR800 UV is equipped with four sources (488nm 532nm, 660nm, 785nm). They are polarised, enabling measurement of depolarisation ratios and studies of orientation within materials. The light is projected to a diffraction limited spot of ~1μm via a x100 water immersion objective lens of an Olympus microscope (other objectives available on the microscope carousel include x10, x20, and x50). The backscattered light is collected by the objective lens in a confocal geometry, and is dispersed onto an air cooled CCD detector (1024x256 pixels) by one of two interchangeable gratings, 1800 lines/mm or 600 lines/mm in the case of the Labram 1B and 900 lines/mm or 300 lines/mm for the Labram HR800 UV, allowing the range from $150cm^{-1}$ to $4000cm^{-1}$ to be covered in a combination of measurement windows. The dispersion of the system operating with the 1800 lines/mm at 514.5 nm excitation (Labram 1B) is ~$1.65cm^{-1}$/pixel with an intensity resolution of 16 bit per pixel and therefore a maximum count of 65535 units. The confocal microscopic system allows the measurement of powdered samples with no further sample preparation, and direct measurement of solids, liquids and solutions. Spectral X-Y Raman mapping can be performed and using the confocality option, a Z stack can be measured. The systems can be equipped with a remote head (with the use of a fibre optic extension) coupled to the spectrometer. Figure 2-3 shows a photograph of the Horiba Jobin Yvon Labram HR800 UV system.

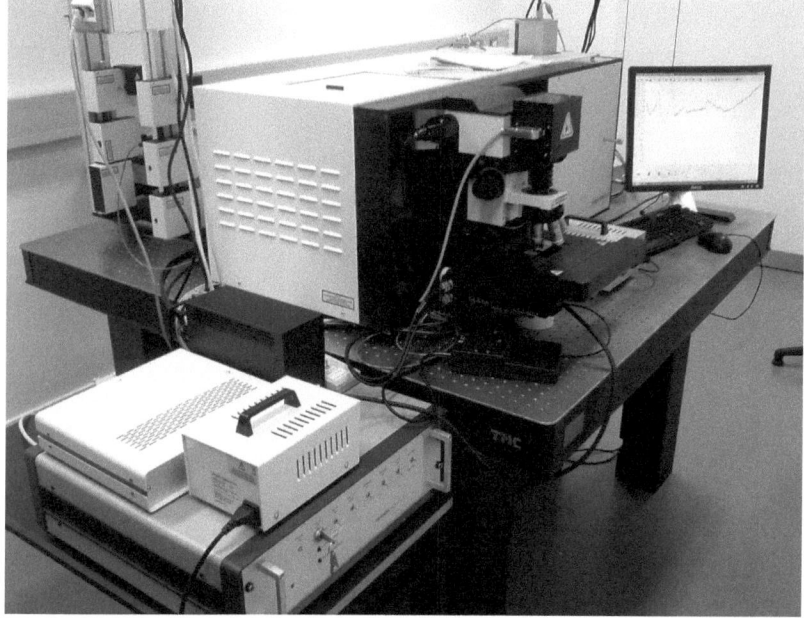

Figure 2-3 Image of Horiba Jobin Yvon SA Labram Raman Spectrometer HR 800 UV

2.4 Materials & Methods

2.4.1 Substrate

The choice of substrate for Raman spectroscopy of *in vitro* cell cultures is a crucial consideration in the experimental design. On the one hand, the substrate should be biocompatible in order to achieve good attachment of the cells. On the other hand, the substrate should ideally contribute negligible or at least well defined at the applied excitation wavelength, thus substrate spectra can vary with a changed excitation wavelength, and therefore contribute with a de-convolvable Raman spectrum. Common culture substrates such as glass slides and transparent plastics produce either undesirable spectral peaks or variable spectra, while slides made of quartz have been shown to give rise to a constant and mathematically removable spectrum [109]. However, cell culture directly on the spectroscopic substrate has proven to influence the chemical content of the cell and its morphology [91, 109].

Even the surface morphology can influence the cell adherence, and culture on biocompatible coatings, such as gelatine or specifically treated surfaces is required. Such coatings have been proven to have negligible influence on the Raman spectrum itself [91]. For *in vitro* studies, a potential candidate biocompatible coating for the cell culture model ideally should reflect the characteristics of the targeted tissue that is supposed to be mimicked. Recent work suggests that the use of 3D-collagen gels outperforms other substrates [110]. However, this being introduced after the experiments in this work, gelatine coated or polished quartz substrates were used throughout.

2.4.2 Cell line

A549 cells (CCL-185, American Type Culture Collection, Manassas, VA), a Caucasian human lung carcinoma epithelial cell line, with dimensions between 35-50 µm, were employed for this study (Figure 2-4). The A549 cell type is used as a model for primary human pulmonary alveolar type II (ATII) cells of the pulmonal epithelial barrier. This cell type secretes a choline based phospholipid lung surfactant (e.g. Dipalmitoylphosphatidylcholine (DPPC)), stored in the membrane bound organelles termed lamellar bodies or keratinosomes [111]. ATII type cells are able to differentiate spontaneously into type I cells within 1-2 weeks [112]. A549 cells are commonly used as a model for pulmonary toxicity assessment induced by biochemical toxins [113, 114]. However recent studies indicate that A549 cells are a better model for alveolar type I cells [112].

ATCC Number: **CCL-185**
Designation: **A-549**

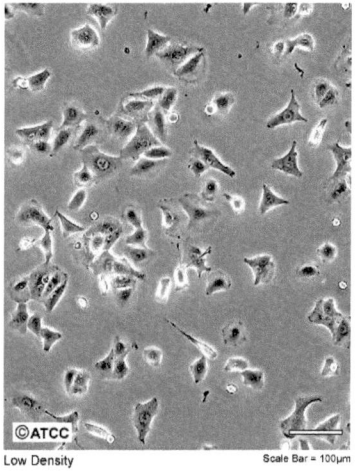

Low Density   Scale Bar = 100µm

Figure 2-4 Microscopic Image of CCL-185 A549 human alveolar, cancerous lung cells of the American type culture collection.

2.4.3 Cell Culture

The cells were cultured in Dulbecco's modified minimum essential medium. (DMEM, Cambrex) The medium contains a number of components including amino acids, vitamins, phenol red and was supplemented with 10% foetal bovine serum (FBS), 45 µg /ml penicillin and 45 µg / ml streptomycin at 37°C in a 5% CO_2 humidified incubator.

2.4.4 Nano-particles

HiPco® derived SWCNT were purchased from Carbon Nanotechnologies, Inc. (Houston, TX). This material contained ~10-wt % iron, as specified by the manufacturer. The diameter distribution of these HiPco® tubes was previously determined to be 0.8-1.2 nm via Raman spectroscopy by calculation, employing the RBM profiles of the SWCNT mixture [115], and TEM [13] in our laboratory. This material, unpurified, has previously been used in parallel studies for colorimetric and clonogenic assays to assess their toxicity to biological material [52, 75, 77, 102].

2.4.5 Exposure

Multiple samples were prepared by harvesting approximately 2.5×10^4 - 2×10^6 cells, depending on the experiment. They were incubated for 24 to 48 h on either gelatine coated or polished quartz slides to facilitate attachment to the substrate. Afterwards, the unattached cells were rinsed off with phosphate buffered saline (PBS) and the exposure agent suspensions were applied to the samples for up to 96 hours, depending on the experiment. The exposure strategy for the direct exposure experiments matched exactly that used by Herzog et al. [99]. The exposure strategy for the medium depletion study matched exactly the one used by Casey et al. [77, 101]. Exposure concentration, time intervals, and preparation parameters were kept similar to deliver exactly comparable data. After exposure, the agent suspensions were then thoroughly rinsed off with PBS and subsequently the exposed cells were fixed in 4% formalin in PBS solution for 10 minutes. As the final step, the samples were washed three times with deionized water (dH_2O). Until spectroscopic measurements were taken, the samples were stored in dH_2O at 4°C in Para film© sealed six well plates.

2.4.6 Experimental measurement setup

Raman Spectroscopy was carried out using either a 514.5 nm (Argon ion) or 532 nm (frequency doubled Nd^{3+}:YAG) laser source and a grating of 300 or 1800 lines/mm, providing a spectral dispersion of about $1.43 cm^{-1}$ or $1\ cm^{-1}$ per pixel (Labram HR 800) . The spectra were recorded using a water immersion lens (Olympus Lum-Plan FL x100) from substrates immersed in dH_2O in a sealed immersion vessel to avoid desiccation of the samples and to prevent possible overheating of the sample. The immersion reservoir was constructed by inserting a quartz window into the bottom of a Petri dish. The x100 water immersion objective gave a spatial resolution of approximately 1μm at the sample. All recorded spectra were acquired as an average of three repetitive measurements at one point to reduce the influence of spectral noise. The system was previously calibrated to the phonon of crystalline silicon, at $520.7\ cm^{-1}$, and depending on the experiment, intensity corrected [116]. The measurements were taken at a constant room temperature of 21°C to match the optimum operating range of the spectrometer. The measurement range was set to an interval of ~250-1750 cm^{-1} with respect to the excitation line. This area covers the fingerprint region of biological samples [76] and in the case of SWCNT as external agent, many of the characteristic SWCNT spectral features [102]. After a series of spectral measurements on a particular slide, the spectral background of the substrate was acquired for reference. The laser power was set to 23-37 mW, depending on experiment and spectrometer, at the sample and the integration time was set to 90 s throughout which delivered reasonable spectra.

2.4.7 Data Analysis platform

The data were recorded with NGS-Labspec version 5 and individually stored in the proprietary Labspec file format to maintain full information content. Every step of the data analysis was then performed using MATLAB (MathWorks, USA) version R2008b and in-house code developments, supported by a High Performance multicore distributed computing cluster of 12 nodes equivalent to 12 personal computers with the total calculatory capability of ~3.5×10^5 million Instructions per second [MIPS].

2.5 Data Processing

2.5.1 Data Preprocessing

2.5.1.1 General Considerations

In Raman spectroscopy, data processing plays an important role in terms of pre- and post- processing. In terms of preprocessing, calibration scales the intensity and frequency axes to a standard, whereas, among other processes, background correction removes unwanted features in the signal [117]. Some of the routines developed and implemented throughout this work are introduced in this section and are attached in the appendix. Thus several approaches were considered, the most promising though was always the one with the least preprocessing and therefore the least impact on the recorded data. Though several methods were developed and scrutinised, from the second experiment on, no filtering and noise correction methods were applied.

2.5.1.2 Intensity Calibration

Intensity calibration is necessary to ensure that the results from different instruments and laser sources are comparable. To this end, the system intensity response was recorded. The use of Standard Reference Material (SRM) No. 2243 of the National Institute of Standards, Boulder, Colorado, USA (NIST SRM 2243, 2242, 2241) provides a means to correct Raman spectra for relative intensity on a day-to-day basis. The application of this standard requires measurements of its luminescence spectrum on the Raman instrument employed. Subsequent mathematical treatment of both the observed luminescence spectrum of the intensity standard and the observed Raman spectrum of the measured sample create the intended comparability of data between spectrometer and excitation lines. The relative intensities of measured Raman spectra are corrected for instrument specific response employing computational methods using a correction curve. These curves are generated with certified polynomials and pre-recorded fluorescence spectra of the SRM glass (a manganese doped borate matrix glass). Accordingly, the spectral range of certification which covers the Stokes area of the Raman spectrum between 200 and 4800 cm^{-1} can be corrected to relative spectral intensity with the following polynomial (Equation 2-8).

$$I_{SRM}(\Delta\nu) = A_0 + A_1 \times (\Delta\nu)^1 + A_2 \times (\Delta\nu)^2 + A_3 \times (\Delta\nu)^3 + A_4 \times (\Delta\nu)^4 + A_5 \times (\Delta\nu)^5 + A_6 \times (\Delta\nu)^6 \qquad \text{Equation 2-8}$$

where ($\Delta\nu$) relates to wavenumbers in cm^{-1} and A_n represents coefficients listed in Table 2-1 (e.g. for 514.5 nm excitation wavelength).

Table 2-1 Coefficients of the certified polynomial for 514.5 nm intensity calibration

Polynomial Coefficient	Certified Value Polynomial Coefficient 20 to 25°C	Polynomial Coefficient of the relative +/- 2σ confidence Curves	
		+2 σ CC	-2 σ CC
A_0	-0.0244612	0.0284858	-0.0284858
A_1	3.17690E-04	-3.17886E-05	3.17886E-05
A_2	-4.84706E-07	1.08168E-08	-1.08168E-08
A_3	4.90077E-10	-1.49980E-12	1.49980E-12
A_4	-1.70340E-13	1.93433E-15	-1.93433E-15
A_5	2.38545E-17	-7.06238E-19	7.06238E-19
A_6	-1.16921E-21	6.87758E-23	-6.87758E-23

By applying Equation 2-8 to the corresponding spectral window used for the acquisition of the luminescence spectrum of the SRM and the sample, the elements of $I_{SRM}(\Delta\nu)$ are obtained. $I_{SRM}(\Delta\nu)$ is normalized to unity over the spectral window and is expressed in terms of photons sec^{-1} cm^{-2} (cm^{-1})$^{-1}$. For the intended application of the luminescence standard, the measured datasets of the SRM luminescence spectrum, $S_{SRM}(\Delta\upsilon)$, and the measured Raman Spectra of the samples, $S_{MEAS}(\Delta\upsilon)$, have to be presented in units of Raman shift (cm^{-1}). The correction curve $I_{CORR}(\Delta\upsilon)$ is defined by Equation 2-9. Its elements are obtained from $I_{SRM}(\Delta\upsilon)$ and of the glass luminescence spectrum $S_{SRM}(\Delta\upsilon)$.

$$I_{CORR}(\Delta\upsilon) = \frac{I_{SRM}(\Delta\upsilon)}{S_{SRM}(\Delta\upsilon)}$$ Equation 2-9

The data points of the intensity corrected Raman spectrum $S_{CORR}(\Delta\upsilon)$ are calculated by multiplication of the data points of the measured Raman spectrum, $S_{MEAS}(\Delta\upsilon)$ of the sample, with the elements of the correction curve as described by Equation 2-10.

$$S_{CORR}(\Delta\upsilon) = S_{MEAS}(\Delta\upsilon) \times I_{CORR}(\Delta\upsilon)$$ Equation 2-10

In measurements employing a dispersive spectrometer, the units of the x-axis (cm^{-1}) are directly related to the wavelength (nm) of the measured spectrum. The calculated spectrum of the SRM for 514.5nm excitation results in the distinctive curvature (Figure 2-5) and with the difference to the measured spectrum the calibration can be achieved.

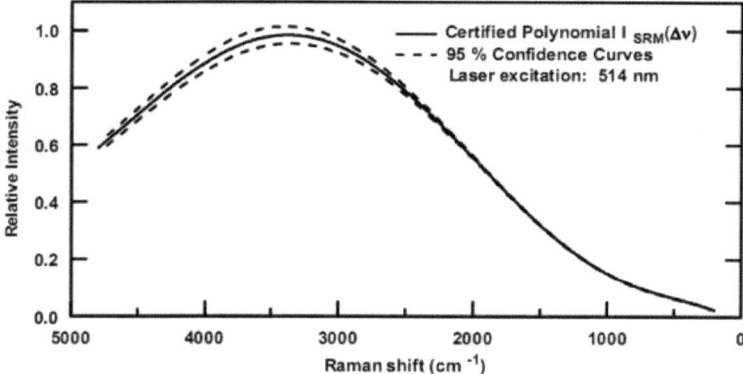

Figure 2-5 Sixth-order certified polynomial for SRM 2243. The x-axis is expressed in Raman shift (cm^{-1}) relative to 514.5 (19435.18 cm^{-1})

Due to the lack of noise in the calculated reference spectrum, the noise of the measured calibration spectrum is consequently conveyed into each, in this manner, calibrated spectrum. Therefore it was found unsuitable after the first experiment. Due to the inhomogeneity in application of existing references, and the capability of the induction of massive noise, this calibration method was only used in the direct exposure experiment.

2.5.1.3 Linearization

In dispersive multichannel Raman spectrometers employing a CCD sensor, the data point spacing during acquisition can be irregular not only due to the use of different gratings. It may change from day to day due to different calibration settings. This leads to recordings with different abscissa and abscissa-linearity therefore induces variance in the x-axes which can be understood as x-noise in the x-direction as described later. Figure 2-6 shows the oscillation of the abscisse-interval of a measured spectrum, possibly as a result of rounding errors induced by the limitation of the spectrometer and the latent under sampling of the CCD due to its digital character. The distance of each datapoint to each neighbour in the x-direction varies around 1.425 cm^{-1} ±.025 cm^{-1}, dependent on the spectral resolution employed. For successful calibration and continuity of the spectral data, this influence can be overcome by interpolation between each data point and subsequent re-sampling of the spectrum. Figure 2-7 shows the linearised spectrum with constant abscissa interval.

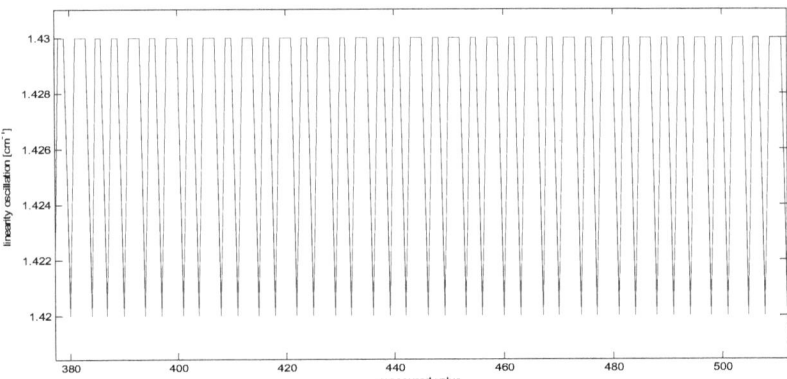

Figure 2-6 oscillation of abscissa-interval of a recorded signal

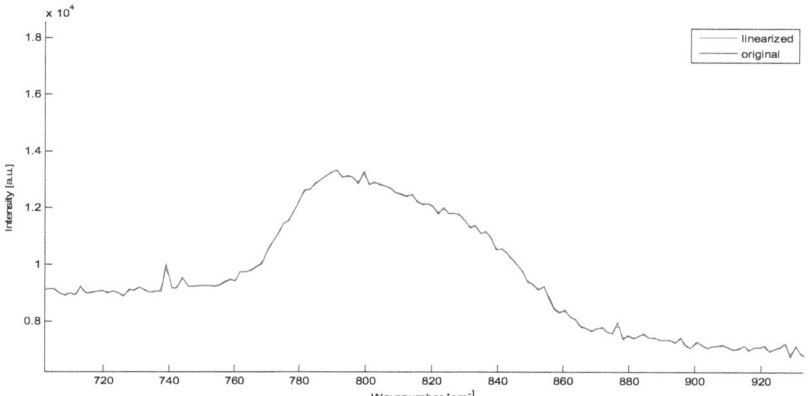

Figure 2-7 Zoomed spectrum of a quartz substrate background, original spectrum (blue), linearized spectrum (red)

Although it is difficult to notice the linearization change even when zoomed in, the linearization leads to a loss in variance in x-direction.

2.5.1.4 Cross correlation offset correction

Offsets in day to day calibrations and possible laser drift over the time period of the measurements can cause a varying shift in the recorded spectra that has to be compensated for in order to reduce the variance along the abscissae. In signal processing, cross-correlation gives a measure of the overlap of two signals [118, 119]. Essentially the compared signals are systematically offset from each other, whilst calculating their product for each offset, which can be employed for Raman spectroscopy, to detect linear shifts in between samples. The highest intensity of the cross correlation function between two spectra denotes their shift relative to each other. Ideally, this shift or lag should be close to zero in unshifted signals.

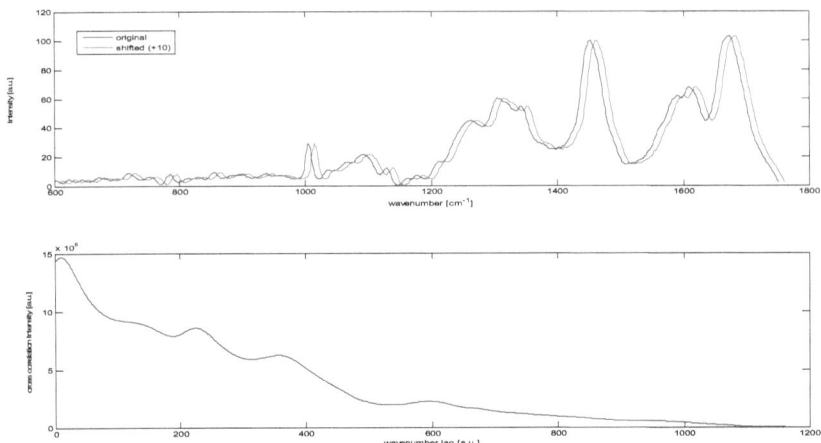

Figure 2-8 Shift detection via cross correlation (shifted spectrum (red) +10, top), cross-correlation intensity maximum at 10, bottom.

Figure 2-8 shows the artificially shifted spectrum of an A549 cell, and the cross correlation results, identifying the induced shift with the maximum at a lag of 10.

2.5.1.5 Noise reduction

2.5.1.5.1 General considerations

Generally, a recorded spectrum can be understood as the sum of the desired and the undesired contributions to the data. The desired contributions should emerge after appropriate preprocessing, and the unwanted contributions such as baseline, substrate or electronic noise should be removed. Noise in general can be used to describe undesired contributions, i.e. the two dimensional variation in intensity in the y-direction and an induced variation in the x-direction.

In theory, a spectrum can be described by a given vector y with i observations of spectral intensities (Equation 2-11),

$$y = \{y_1, y_2, ..., y_i\}$$
Equation 2-11

and it is possible to model it by the sum of an ideal spectrum s and the background b convolved by a blurring function p, any applicable function, which basically is a function that induces various shifts or broadening of features along the x axis, plus the added noise n which *only influences the y axis*. (Equation 2-12). The convolution is denoted by * [117]. This is an extension of the standard model generally used in signal processing [120].

$$y = (s+b)*p+n \qquad \text{Equation 2-12}$$

It is clear that, without additional knowledge, Equation 2-12 is difficult to solve for the desired spectrum. The noise n is usually assumed to be statistically constant and thus its mean contributes to the baseline and its variance constitutes the noise [117, 121]. Usually one can get an impression of the baseline by recording without a sample in place (y_b) but this results in a similarly complex function, Equation 2-13.

$$y_b = (s_b + b_b)*p_b + n_b \qquad \text{Equation 2-13}$$
$$y_g = (s + y_b)*p_b + n_b$$
$$y_g = (s + [(s_b + b_b)*p_b + n_b])*p_g + n_g$$

In an ideal situation, only the signal of the baseline s_b might be free of any additional baseline (b_b=0), no blurring function would exist (p_b=[ ]= empty) and no noise (n_b=0). Thus the blurring function of the ideal spectrum (y_g) would become (p_g), - the baseline itself (Equation 2-14) therefore showing the relation to the standard model.

$$y_g = s + p_g + n_g \qquad \text{Equation 2-14}$$
$$s = y_g - p_g - n_g$$

This spectral model should be solved intrinsically, ideally with the subtraction of elements that are neither signal nor noise. Common practice is to subtract recorded reference spectra of major background contributors (e.g. substrate, dark-current, intensity response, medium) following the principle of superimposition [122]. But also in this case, each spectrum follows the same equation (Equation 2-11) and unfortunately the recorded correction spectra are not free of noise. Therefore, with every subtraction, the noise, which is not stationary or, of predictable intensity, is introduced into the resulting signal. Alternatively, the complex noise can be understood as a blurring function itself, shifting peak positions on sharp edged signals. Therefore, these two kinds of noise, as mentioned previously, have to be overcome. Electronic noise, consisting of flicker noise, shot noise and thermal noise is an unpredictable and constant occurrence primarily in the y-direction, and can have a huge impact on the quality of any signal [123]. To reduce the influence of electronic noise, it is important to increase the signal to noise ratio in the experimental setup itself. An effective way to reduce the electronic noise of a spectral signal is the exploitation of the full detection range of the CCD sensor, by either increasing the collection time or excitation energy. The laser beam has to be focused precisely on the point of interest and the spectra of this point have to be recorded multiple times to statistically remove random noise. Nevertheless, every electronic element in the data collection chain, even the resolution of the A/D converter behind the CCD sensor and the calculation precision of the (commonly used) computer system adds noise, or uncertainties, to the real signal. In this work, the variations in the y-direction are predominantly addressed as noise whereas the variations in the x-direction are addressed by substrate and baseline removal.

2.5.1.5.2 Moving Average and Savitzky-Golay filtering

Smoothing of data is commonly employed to reduce the electronic noise contributions to a signal. Linear moving average filters (MA) (Equation 2-15) are among the simplest smoothing filters.

$$y_{i,new} = \sum_{j=-p}^{p} c_j y_{i+j}$$

Equation 2-15

Over the window of [-p, p] data points, the signal is averaged. In the case of a linear moving average filter, $c_j = 1/(2p-1)$ is constant. The number of points in the filter or filter interval is often referred to as the filter window. The more points used in the filter window, the higher the noise reduction becomes, but at the same time the probability of blurring (distorting) the signal increases. With linearity of the filter comes the disadvantage of approximation of signals. Signal peaks are better approximated by curves, and in a linear model the peak maximum will always be underestimated [120]. Models of higher order usually provide a better approximation. Nonlinear approximated moving average filters are the extension to the classical moving average filters. Each window [-p p] is regressed with a function of higher order and the central value of this window is replaced by the closest estimate. This tedious process was simplified by Savitzky & Golay in 1964 [124], by expressing these regression calculations for a certain window and order as a sum of coefficients also referred as weights. For example, in a quadratic SG-filter with seven points, the coefficients c = [-2, 3, 6, 7, 6, 3, -2]. Therefore the first SG-filtered datapoint will be calculated according to Equation 2-16.

$$\hat{y}_{i,sg} = \sum_{j=-p}^{p}(c_j / \sum c) \cdot y_{i+j}$$

Equation 2-16

$$\hat{y}_{1,sg} = \begin{bmatrix} -(2/21) \cdot y_1 \\ +(3/21) \cdot y_2 \\ +(6/21) \cdot y_3 \\ +(7/21) \cdot y_4 \\ +(6/21) \cdot y_5 \\ +(7/21) \cdot y_6 \\ -(2/21) \cdot y_7 \end{bmatrix}$$

Although this moving average-filtering involves higher order approximations, it is still a linear model. Each filtered point is a linear combination of the original data. Nevertheless, the Savitzky-Golay smoothing algorithm has proven to be useful for reducing the noise levels of spectroscopic data [125]. However, it is less effective in cutting out noise than normal finite impulse response filters (FIR). A problem is the preservation of high frequency impulses that can be noise [126] and the possibility of induced variance by the far window weight on the local, to be smoothed, signal value. Therefore, it is possible to induce shifts of peaks of low intensity. Figure 2-9 shows the influence of order and window size of Savitzky-Golay filters for noise reduction used in literature [91, 127-129]. One can easily observe that sharp features broaden and decrease with increased window size, (see Phenylalanine peak at 1003 cm^{-1}) whereas the peak position is influenced by the combination of order and window size. Therefore both effects can induce additional variance in a multivariate approach.

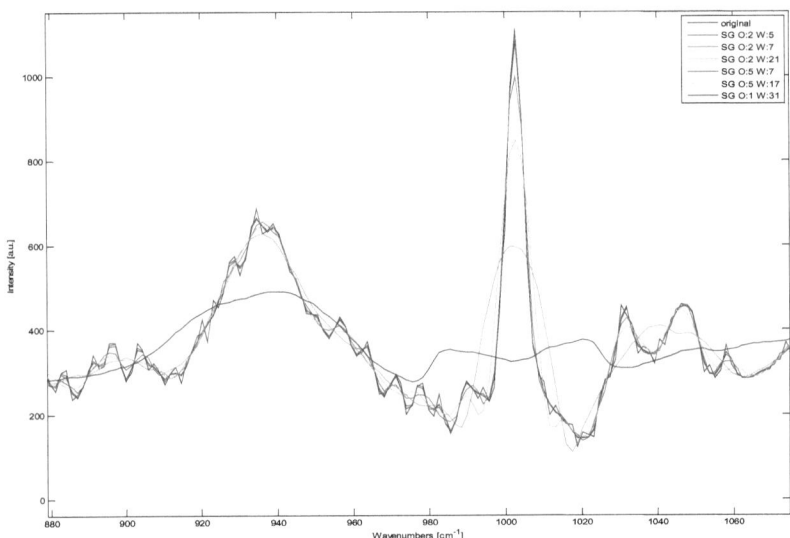

Figure 2-9 Influence of S/G filter settings used in literature on a cellular spectrum of A549 in the region 970cm-1-1070 cm-1

2.5.1.5.3 Moving Median Smoothing

The running median smoothing (RMS) approach is likely to smooth features of the distribution in a data window, whilst rejecting extreme outliers. The window size determines the degree of smoothing, similar to the moving average filter. The broader the window, the more features removed and their intensity and slew rate drops. Figure 2-10 shows the influence of different running median filters on a sample spectrum.

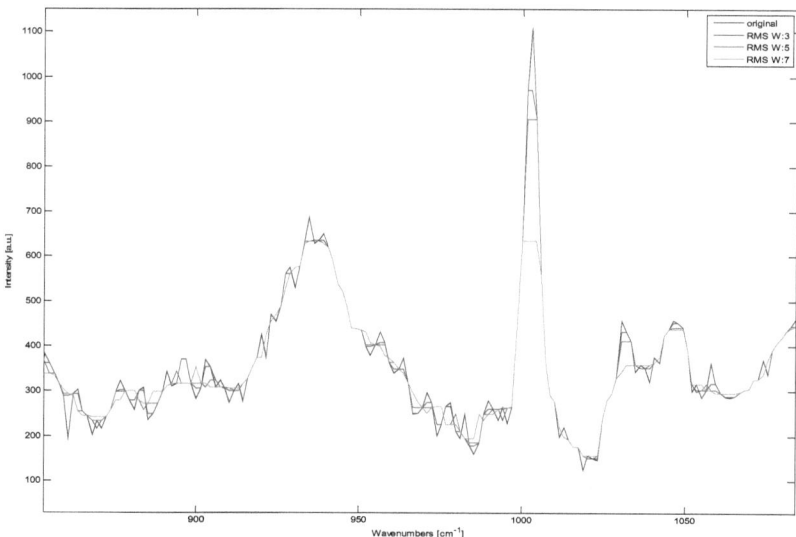

Figure 2-10 Influence of RMS filter settings on a cellular spectrum of A549 in the region 970cm^{-1} - 1070 cm^{-1}

With increasing window size, the intensity of the sharp features (e.g. Phenylalanine) is reduced, whereas small, rather broad features (e.g. C-H in plane bending at 1033cm^{-1} and the C-O stretching at 1046cm^{-1}), are not blurred as can be seen with a MA smoothing with a narrower window of seven points (Figure 2-9). In regions of the spectrum where the feature intensity comes close to the noise level, the information of the x-location becomes blurred. In direct comparison, the RMS filter is closer to the real signal whereas with the moving average filter the definition of the phenylalanine peak is lost (its base widens) (Figure 2-11).

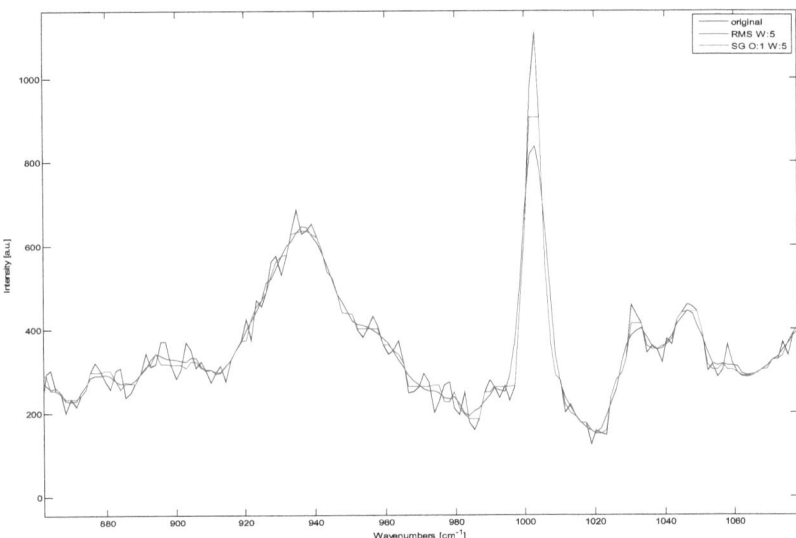

Figure 2-11 Comparison between SG, MA (1st order window 5 points) and RMS (window 5 points) filters on a cellular spectrum of A549 in the region 970cm^{-1} - 1070 cm^{-1}

An alternative, self developed, way of employing a running median smoothing function with improved performance is to focus on a RMS with a three point window. As Figure 2-10 shows, the RMS with a window size 3 or 5 comes closest to the original signal and rejects a substantial amount of noise. Assuming, on a very noisy signal, a 5 point window can be understood as two 3 point windows shifted by two points, the median between these five points can be, in a worst case scenario, described as the median between only local maxima and local minima of five points. By extending this scheme, the ideal median can then be understood as the median between the local maxima and local minima only, because a curve (or signal) of any kind is described predominantly by its minima and maxima [130]. This assumption was adopted to construct a different median based filter that interpolates between all local minima, the bottom margin of trust region, and all local maxima, the top margin of trust region, independently over the whole abscisse of the spectrum. Therefore it returns the median between both. It builds on the assumption that the signal to noise ratio is high in areas of signal which narrows the distance between bottom and top margin of trust region due to the lack of noise spikes. Therefore it doesn't influence the actual signal features but removes the noise only (Figure 2-12).

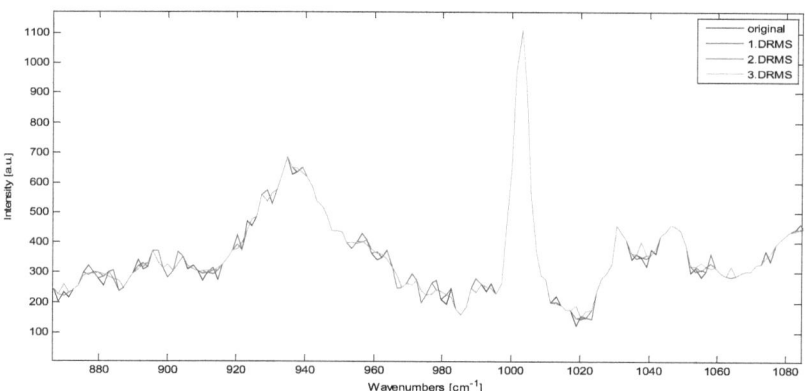

Figure 2-12 DRMS Filter applied to an A549 spectrum between 870 cm^{-1} - 1070cm^{-1}, original (blue),1.run (green), 2.run (red), 3.run (cyan)

It basically takes advantage of the signal dynamics (i.e. sharpness of the features) and is therefore termed Dynamics-assisted RMS (DRMS). Every time the signal is passed through this filter it estimates the reliable area of a signal and calculates a median through this band. The first and second iteration give a reasonably good result, with good noise removal, while preserving all features and not inducing any broadening or flattening. The third pass induces some broadening by noise in certain low frequency areas (1036 cm^{-1}) but still preserves the dominant features to their full extent.

The sum of the absolute residuals after subtraction of the smoothed from the original signal is shown in

Table 2-2.for variations of all methods explored. The smallest absolute residual indicates the best filter performance. Therefore the DRMS is expected to deliver the best performance, by returning the most realistic spectrum after noise filtering.

Table 2-2 Noise Reduction Performance of common and proprietary noise filters

Smoothing Filter	Smoothing Parameter	Σ(Absolute(Residual))
MA-SG	order : 1 window : 5	22,268
MA-SG	order : 1 window : 31	59,435
MA-SG	order : 2 window : 5	12,645
MA-SG	order : 2 window : 7	17,872
MA-SG	order : 2 window : 21	29,630
MA-SG	order : 5 window : 7	10,927
MA-SG	order : 5 window : 17	22,267
RMS	Window : 3	9,993
RMS	Window : 5	16,113
RMS	Window : 7	19,344
DRMS	pass : 1	6,390
DRMS	pass : 2	8,621
DRMS	pass : 3	12,788

2.5.1.5.4 Histogram filtering or data binning

Histogram based filtering, also known as data binning, is a technique also used for data pre-processing and noise reduction [131]. Assuming that the unwanted signal, or noise, is only of a sinusoidal nature, it can be filtered with a low-pass filter [132] or each peak on top of the baseline can be binned and replaced by its mean as shown in the simulated dataset of Figure 2-13. Both strategies would remove the artificial recurring noise equally well. The histogram as such shows in this setup a very dense distribution of 23 nearly similar area occurrences, indicating the artificial nature of the simulated dataset, a single overlayed sine function.

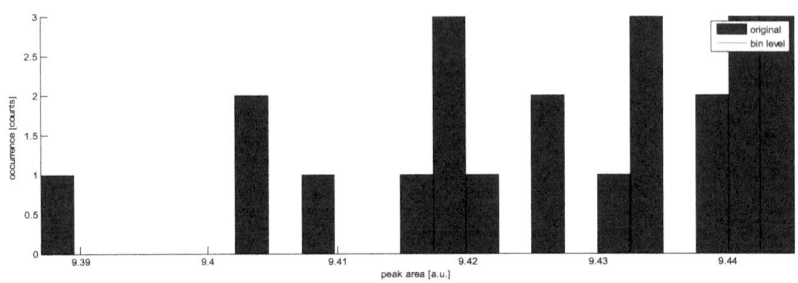

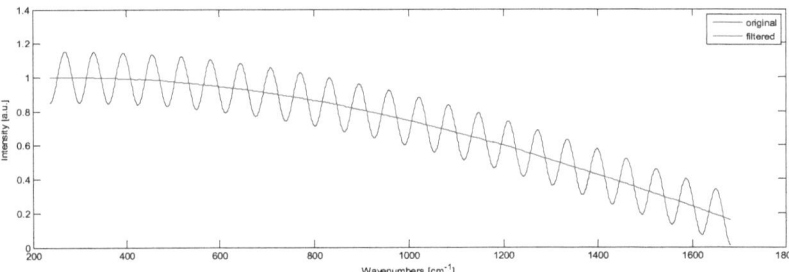

Figure 2-13 Histogram and binning result (binned features =23, outside of the histogram) of a baseline overlaid with a sine function representing noise

In Raman Spectroscopy, the noise is usually not entirely of a systematic nature. Therefore a different binning algorithm was developed that bins signal features according to their area as a measure of the power of the feature. It is assumed that every feature is superimposed on a baseline that is described by an interpolation between all local minima of the spectrum. The number of bins is then calculated from the number of similar area peaks. Further, it is assumed that the signal part is significantly different to the noise. Signal is understood as broad and high in amplitude but relatively rare, whereas noise is understood as sharp and shallow, but of frequent occurrence. Therefore, a histogram of a real spectrum should show an exponential decline in occurrence of peak area sizes as seen in Figure 2-14. The high frequency noise contribution is represented on the left and the low frequent signal contribution on the right.

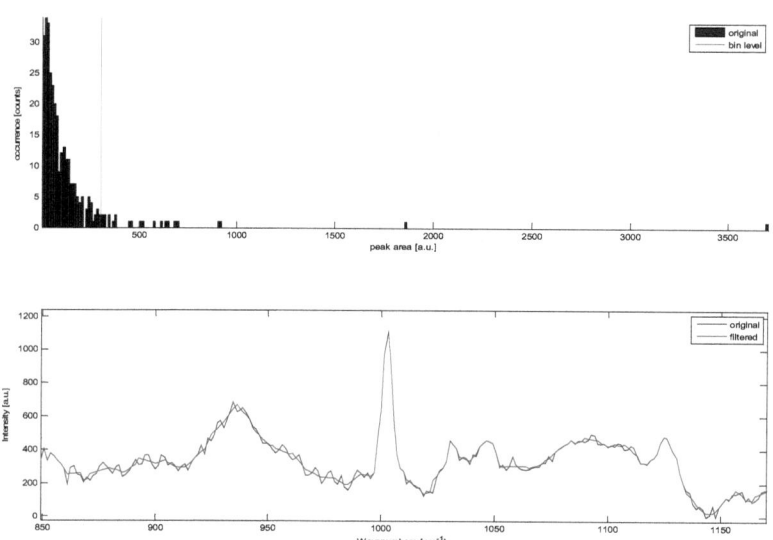

Figure 2-14 Histogram and binning result (binned features = 300) of a spectrum of A549 (850cm^{-1} - 1170cm^{-1})

Depending on the binning level, the spectral features are not distorted and an absolute residual in this example can be calculated to be 18.556. Thus, this method is comparable to the performance of the MA & RMS filter techniques

Table 2-2 whilst removing a higher amount of noise than the best DRMS. As a rule of thumb it was found that the best binning level (rejected peak area above baseline) is correlated to the sharpest feature that has to be resolved (e.g. Phenyl ring breathing mode at $1003cm^{-1}$ for biological samples in Raman Spectroscopy). Therefore, a combination of a single run DRMS filter and a histogram binning filter for analysis is proposed instead of MA filters in the case that the recorded and corrected spectra are very noisy, due to the influence of the applied background and baseline correction (shown later).

2.5.1.6 Baseline correction methods

2.5.1.6.1 General considerations

In Raman spectroscopy, recorded spectra usually contain chemical features and others which originate in spurious or constant physical effects occurring during the measurement. These features are usually of low-frequency, broad and are generally described as 'baseline' and/or 'back-ground' [117]. They are commonly removed by one of numerous algorithms [133, 134] e.g. one automated approach is the implementation of the so-called EMSC (extended multiplicative scatter correction) algorithm, originally developed for IR spectroscopy [135]. Common semi-automated computational methods of background subtraction are the subtraction of a polynomial of certain order [136-138], a slope and/or an offset fitted to each spectrum. The use of derivatized data instantly removes the baseline and increases the spectral resolution, but the amount of sharp noise related features is duplicated and the information gained is more difficult to interpret [139]. In this work, the baseline removal strategies focus on the integration of previous knowledge, separating the substrate and the baseline and its effects on the signal as such are treated separately.

2.5.1.6.2 Manual baseline removal

In manual baseline removal, commonly the baseline is removed by a semiautomatic subtraction of a manually selected set of nodes that support the baseline. Certain points of background contributed nodes in the measurement window that apply to every spectrum are defined. These points are then interpolated for each spectrum individually by a linear or higher order function and then subtracted. This strategy is based on the nonnegative subtraction of the integrated area described by the spectral curve with the trapezoidal rule [140]. It is strongly dependent on the node selection and the knowledge of the curvature of the baseline of a spectrum. It is very labour intensive on large datasets, due to the requirement of individually processing of each spectrum and can interfere with the substrate removal. Manual baseline removal induces signal variance between sample and measurement. It is subjective and is not explicitly suitable for automated baseline removal or comparison [121, 139].

2.5.1.6.3 Rubberband baseline removal

This baseline removal strategy addresses the influences of the spectral features outside the recording window that leak into the recorded spectrum (e.g. Rayleigh scattering [141]). It was programmed for the Matlab environment, mimicking an available, but here automatic, preprocessing option in the commercially available "Opus" software package commonly used on Bruker FTIR spectrometers. This function stretches linear segments of baseline between the local minima of the signal along the spectrum, starting from the lowest local minima to both endpoints of the spectral window, while always staying below or equal to the spectral value. In the case where the slope of the stretched segment does not touch the signal at the next node, the slope is automatically increased until it matches. It is capable of removing linear to concave baseline contributions. It was preferred over the squared exponential baseline subtraction suggested by Bulmer et al.[133], due to its ability to exactly fit the real data on both sides of the spectrum with a line shape that can be anything between linear and exponential up to a polynomial of second order.

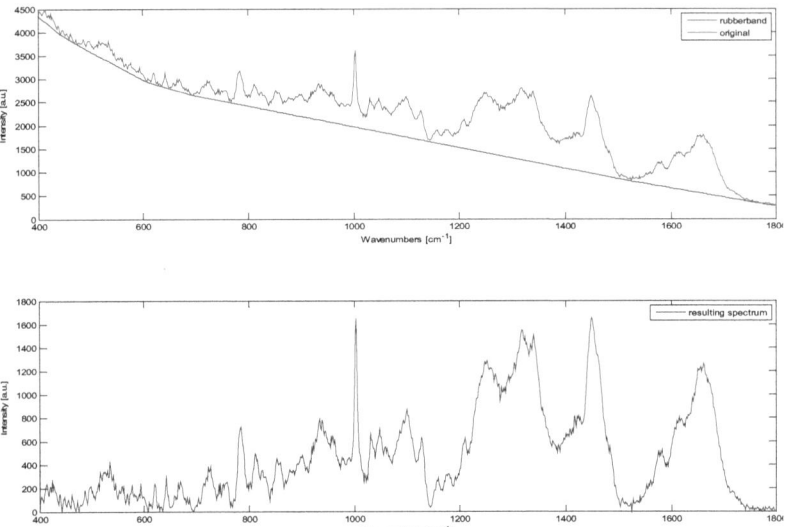

Figure 2-15 Baseline removal with the rubberband method. Original spectrum (top), baseline removed spectrum (bottom)

In Figure 2-15, the application of the rubberband baseline removal algorithm to a spectrum of an A549 cell is demonstrated. The signal to noise ratio (SNR) given by Equation 2-17 [142] can be used to estimate the improvement (gain) of the signal after processing. For automatic SNR estimation, the signal (S) used is the strongest spectral feature between local minima, whereas the noise (N) is represented by the smallest.

$$SNR = \frac{P_S}{P_N} = \left(\frac{A_S}{A_N}\right)^2$$

$$SNR_{dB} = 10 \cdot \log_{10}\left(\frac{P_S}{P_N}\right) = 20 \cdot \log_{10}\left(\frac{A_S}{A_N}\right)$$

Equation 2-17

The SNR_{dB} changes due to rubberband removal are only from 25.78 dB to 25.05 dB, indicating only a tiny increase in noise, which is caused by the reduced power (P_s) of the signal feature in comparison to the power of the noise (P_n). Furthermore, no negative values are created and all features are preserved at the original spectral location. Therefore the rubberband method qualifies as an appropriate method for baseline subtraction.

2.5.1.7 Substrate removal methods

2.5.1.7.1 General considerations

Although in confocal Raman spectroscopy the contribution of the underlying substrate of each sample is minimal, contributions of the substrate are nevertheless omnipresent. Depending on the opacity and thickness of the sample, the amount of backscattered light by the substrate contributes to the background or offset in the spectra differently. Therefore the substrate spectrum, embedded in the analyte spectrum, will alter accordingly. It is common practice to record reference substrate spectra, in order to subtract them from the analyte [91, 93]. In the following section strategies developed over the course of this work are introduced to demonstrate the optimal solution for quantitative and qualitative substrate removal.

2.5.1.7.2 Subtraction of reference background

The subtraction of the reference from the real signal as a substrate removal strategy has two implications. The first is the qualitative implication – for example, are the abscissae of the two spectra completely identical? The second is the scale, the quantitative implication - is the intensity of the substrate spectrum alone relatable to its contribution to the spectrum? Assuming the principle of superimposition, the substrate can be scaled to its contribution before subtraction, as shown in Figure 2-16.

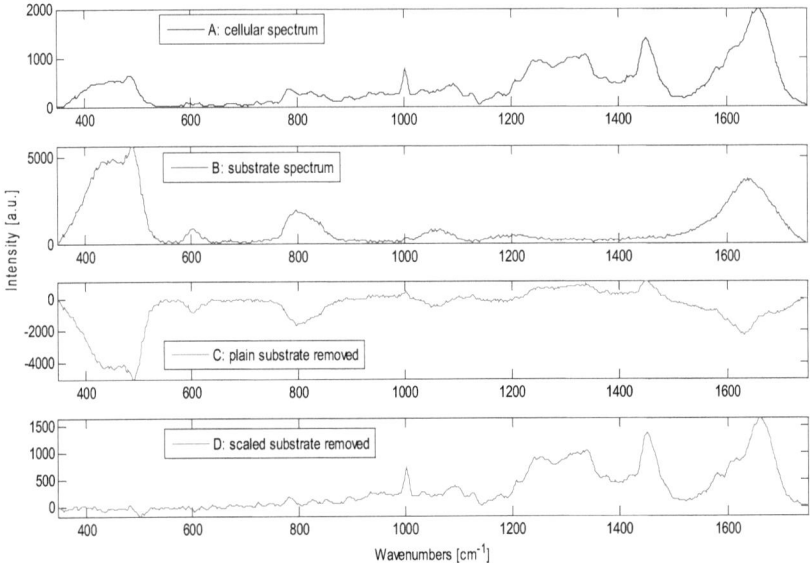

Figure 2-16 Substrate (B) removal by subtraction of a recorded substrate (C, SNR_{dB} 24.88) from a Cellular A549 spectrum(A, SNR_{dB} 21.73), and a recorded substrate scaled to the signal (D, SNR_{dB} 23.79)

The direct subtraction of the recorded quartz spectrum in water immersion from a cellular spectrum of an A549 cell (A-B=C) results in a slightly higher signal to noise ratio of the result, due to the removal of the statistically constant but contrary noise. Unfortunately, a large number of negative features are introduced into the result, which renders the advantage of increased SNR irrelevant. Ratioing, or scaling of the background before subtraction is a common method in spectroscopy [121, 143]. By scaling the substrate intensity to the signal (A-(S*B)=D) the SNR is increased and the amount of negative features induced is significantly lower. The disadvantage is the possibility of altering the intensity of the noise by multiplying it by the scaling factor according to Equation 2-18.

$$s = y_g - n_g - c \cdot (s_s + n_s)$$
$$= y_g - n_g - cs_s - cn_s$$

Equation 2-18

This may shift the scaled noise of the substrate cn_s to a different order of magnitude than the noise of n_g. Therefore this method can induce variance due to reduced auto annihilation of opposing noise variations. This auto annihilation is commonly used in the averaging of measurements during recording. In those circumstances the noise is of equal magnitude therefore auto annihilation occurs. In the case that auto annihilation does not occur the noise adds up and creates spectral features which are still caused by noise. This effect is commonly addressed by extensive filtering of the substrate prior to subtraction. By doing this the disadvantage shown in the noise filtering section earlier, can appear. An ideal approach is subtracting a substrate signal that is noise free [121, 126].

2.5.1.7.3 Subtraction of modelled background

This alternative substrate subtraction method takes the idea of noiseless subtraction onboard and consists of two interlinked parts. The first part is the characterisation of the substrate spectra in order to create a quantitative model. The second part consists of the application of this noise free model to the spectra. To achieve this, the substrate spectra are scaled after baseline removal with the rubberband method to the characteristic Raman feature of the substrate, in this case, quartz, at 486 cm^{-1} [2, 144], by scaling the signal maximum ±5 cm^{-1} around this feature to 100 percent. The spectra that contain significant quartz features are averaged and the peaks are selected by analysing the first, second and third derivative. After identifying 120 peaks of the substrate spectrum, each substrate spectrum (y_g) was fitted with 120 weighted Gaussian and Lorenzian functions (Equation 2-19, representing 99.5% of the substrate variance) simultaneously, using the "trusted regions methods", provided by MatLab's curve fitting toolbox.

Equation 2-19
$$y_g = +g_1 * a_1 * e^{-\left(\frac{x-x_{p_1}}{c_1}\right)^2} + (1-g_1) * \frac{a_1}{1+\frac{4*(x-x_{p_1})^2}{c_1^2}}$$
$$+ \vdots + \vdots$$
$$+ g_n * a_n * e^{-\left(\frac{x-x_{p_n}}{c_n}\right)^2} + (1-g_n) * \frac{a_n}{1+\frac{4*(x-x_{p_n})^2}{c_n^2}}$$

The optimal fit, with a maximal standard deviation of 5% per spectrum, is obtained without any other boundary definition than peak position ($x_{p(n)}$) and weight (g_p) limit (g_n=[0 1.1]) with 10^6 iterations per peak, excluding the H_2O feature at 1643 cm^{-1}, in the case of immersion measurements, returning the width (c_p) and the amplitude (a_p)

and the calculated weight (g_p). The entire fit parameters were then used to rebuild the calculated spectrum of weighted Gaussian and Lorentzian functions. The peak parameters that describe the fit to 99% are employed to build the substrate model in order to reduce the number of parameters. The standard deviation, per parameter, delivers the boundary definition for the final fit - model of the substrate to be subtracted from recorded cellular spectra, measured on top of the modelled substrate, as shown in Figure 2-17. The characterisation parameters thus obtained are given in Table 2-3. Surprisingly, the feature taken for scaling turns out to be shifted by about 3 cm^{-1}.

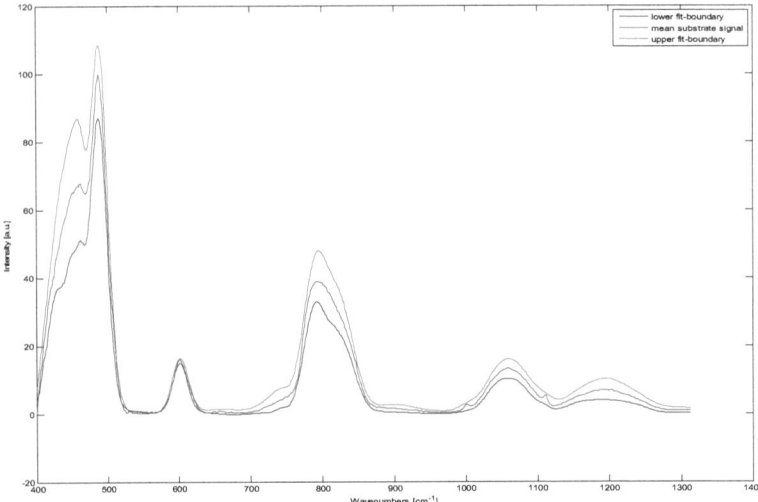

Figure 2-17 Boundaries of the fit parameters above (red) and below (blue) the mean substrate signal (green)

Table 2-3 List of the selected peaks to fit the substrate with their fit parameters.

Peak	upper fitting limits				lower fitting limits				Area
	position	amplitude	width	weight	position	amplitude	width	weight	7871.2
409	409.65	6.01	8.94	1.01	409.32	2.58	3.79	0.99	31.1
426	426.28	31.12	18.34	1.00	426.09	24.55	16.88	1.00	597.9
434	434.37	15.93	20.81	1.00	433.73	7.54	19.43	1.00	292.0
449	448.98	30.81	14.93	1.07	448.94	20.76	11.16	0.81	435.2
459	459.60	5.62	26.39	1.01	458.73	-1.53	17.61	0.72	56.3
462	462.37	24.23	10.87	1.06	462.37	17.89	9.01	0.60	292.3
468	468.10	34.47	34.22	1.10	467.71	25.20	25.94	0.82	1135.0
489	489.44	83.27	16.07	1.00	489.25	72.67	15.27	0.96	1539.7
602	602.30	15.66	16.28	1.08	602.02	14.81	14.26	0.79	301.2
635	635.16	0.70	40.99	0.97	634.55	0.04	24.69	0.66	14.1
727	727.15	1.31	25.19	0.97	726.56	0.14	17.71	0.92	18.8
731	731.57	1.52	22.36	0.98	730.66	0.00	15.91	0.92	17.0
742	742.86	2.27	14.41	1.05	741.91	1.18	10.30	0.83	27.6
749	749.85	1.64	28.68	1.08	748.63	0.17	8.63	0.56	15.6
757	757.31	0.56	36.81	0.98	756.55	-0.08	27.35	0.94	10.2
790	790.03	34.50	23.03	1.00	790.03	28.66	21.91	1.00	882.6
794	794.40	5.20	27.58	1.00	794.33	1.81	24.81	0.99	114.1
807	807.53	3.66	36.32	1.05	806.94	0.07	30.75	0.88	80.2
817	817.48	6.87	21.91	0.76	815.92	4.68	13.05	0.02	190.1
829	829.12	22.24	25.41	1.00	829.10	17.95	23.75	1.00	615.0
880	879.99	0.28	30.22	0.95	879.10	0.09	15.84	0.70	5.7
887	886.96	0.36	29.17	1.03	886.23	0.01	19.11	0.84	5.2
902	902.88	0.39	30.53	0.95	901.98	0.02	25.56	0.86	7.5
910	910.12	0.37	31.43	0.90	909.23	0.06	24.41	0.85	7.7
914	914.35	0.23	32.72	0.88	913.51	0.04	23.74	0.84	5.1
921	921.36	0.31	32.20	0.87	920.64	0.07	21.81	0.82	6.6
1001	1001.02	0.93	18.58	0.98	1000.71	0.17	13.03	0.62	12.2
1004	1003.81	0.73	18.98	0.93	1003.23	0.17	13.36	0.76	9.6
1014	1013.82	0.75	26.26	0.95	1013.40	0.14	20.24	0.92	13.0
1040	1040.14	4.93	23.60	1.00	1039.95	3.53	21.05	0.99	118.5
1050	1050.16	2.07	29.97	0.99	1049.97	1.41	26.58	0.98	61.4
1053	1053.17	1.77	30.66	1.00	1052.82	1.34	28.35	0.97	57.5
1058	1057.90	1.76	31.40	1.00	1057.53	1.25	29.90	0.97	57.9
1062	1061.76	1.74	31.54	1.01	1061.44	1.16	29.90	0.96	56.0
1072	1071.74	2.03	29.41	1.00	1071.72	0.98	26.73	0.98	53.0
1076	1076.02	6.34	27.64	1.01	1075.95	4.54	25.37	0.96	180.7
1108	1108.49	1.87	18.50	0.90	1107.44	1.08	11.26	0.32	34.9
1112	1112.23	1.26	28.31	0.83	1111.76	-0.03	16.99	0.41	17.5
1118	1117.94	0.21	29.90	0.94	1117.55	0.15	26.54	0.85	6.7
1129	1128.88	0.18	28.51	1.01	1128.49	0.11	25.96	0.93	5.2
1135	1135.05	0.31	28.10	1.00	1134.19	0.16	21.21	0.73	7.7
1158	1158.41	3.26	27.46	0.98	1158.21	2.11	23.37	0.95	86.5
1163	1162.71	0.46	28.97	0.97	1162.49	-0.03	25.32	0.94	7.7
1167	1167.04	0.68	30.16	0.97	1166.84	0.16	25.26	0.94	15.1
1170	1169.84	0.84	30.48	0.97	1169.65	0.41	25.78	0.94	22.5
1174	1174.51	0.74	30.20	0.97	1173.99	0.42	25.20	0.87	20.8
1183	1183.75	0.94	28.19	0.99	1182.92	0.57	20.86	0.81	24.0
1188	1188.39	2.25	24.40	1.00	1188.09	1.04	18.61	0.96	42.7
1198	1198.32	2.04	24.63	0.99	1198.11	0.69	21.58	0.89	41.2
1208	1208.25	1.92	27.37	0.96	1207.60	0.86	20.22	0.89	41.1
1212	1212.54	1.41	28.51	0.97	1211.78	0.50	21.06	0.92	29.7
1222	1222.58	0.94	28.62	0.96	1222.27	0.66	24.02	0.92	26.8
1227	1226.92	0.95	27.73	0.99	1226.71	0.60	24.19	0.85	26.1
1231	1231.27	0.94	27.08	0.98	1231.03	0.50	24.01	0.83	24.1
1237	1237.01	1.29	26.14	0.95	1236.75	0.49	21.74	0.89	27.3
1251	1251.00	1.15	25.13	0.92	1250.70	0.57	18.73	0.88	24.2
1269	1269.43	0.67	25.78	0.97	1268.81	-0.09	19.44	0.60	8.9
1311	1311.05	0.80	30.34	0.85	1310.88	-0.06	23.73	0.76	7.1

2.5.1.7.4 Water immersion compensation

In Raman spectroscopy in an immersion setup, the contribution of water to the spectral features of the cell cannot ignored [110]. The vibrational spectrum of water comprises of two major features, the dominant one having origin in the antisymmetric OH stretching vibration at ~3400 cm^{-1}, the minor bending modes occurring at ~ 1640 cm^{-1}. While recording Raman spectra in a confocal configuration, the influence of the environment surrounding the focal spot can be reduced, but due to the relatively thin sample (~ 10µm at the thickest location), the contribution of the immersion liquid and the substrate to the confocal volume becomes considerable, especially during experiments with long acquisition times. To account for these influences, in this work the spectral window in all studies was limited to a maximum of 1750 cm^{-1} in order to prevent detector saturation by the high frequency water features, whilst maintaining the full detector range. The tailing wing of the water feature at 3400 cm^{-1} was found to be minimal in this region and was successfully removed with the above described rubber band method. Only the feature at 1640 cm^{-1}, proved to be a problem, as it is close to the Amide I band at 1656 cm^{-1}, and due to its intensity, can shift its position. Because water can be found inside and outside of the cell, and due to the variance in thickness of the sample at the point of acquisition, the actual amount of water measured is unpredictable. Therefore, it was decided to fit the whole Amide I peak area after substrate and baseline removal with a series of Gaussian and Lorentzian peaks, similar to the process employed in modelling the substrate spectrum, and finally the characteristic feature at 1640 cm^{-1}, originating from the water was subtracted (Figure 2-18).

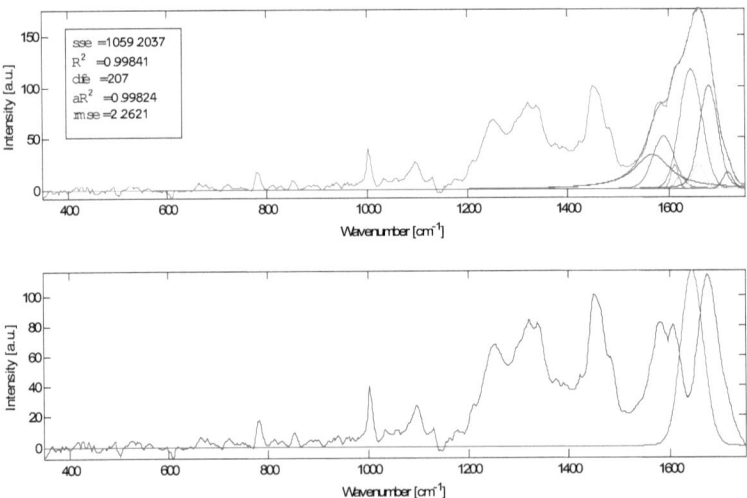

Figure 2-18 Example of the removal of the water feature at ~1640 cm^{-1} from a cellular spectrum of an A549 cell

2.5.2 Univariate Data analysis

2.5.2.1 Spectra Interpretation

In common spectroscopic analyses, the information evident in spectra is extracted by qualitatively analyzing peak intensities, peak areas, peak shapes, shoulder formation and their changes due to external agents (e.g. Amide I band at 1656 cm^{-1}) (

Table 2-4). This is usually done in the derivatized, double derivatized data or the plain spectrum [145, 146]. The peak positions, usually examined, are determined by their correlation to biological functions and metabolic relations. The information so-gained is usually put into context with biological evidence solely, employing references of a number of well defined peak assignments. This univariate approach attempts to address the complex biological reactions of the sample by changes of single or a small number of peak positions and/or intensities (e.g. Perna et al used 1302 cm^{-1}, 1338 cm^{-1}, and the Amide III band for peak ratio analysis) (

Table 2-4). Univariate results (the peak information) are commonly analyzed statistically with respect to a certain observation, in order to explain the supposed effect [91, 107]. Therefore this method focuses on a small number of peaks to analyse biological impact.

Table 2-4 common peak assignments [91, 105, 147]

Wavenumber (cm^{-1})	Assignment
1656 – 1690	Amide I band
	C=O stretching of amide, NH$_2$ in plane bending, C=C stretch (lipids), Amide I (-helix, protein)
1583 – 1605	G-Band of CNT's
1485 – 1660	amino acids, amino acid hydrohalides
~1451	CH$_2$ scissoring v., CH$_3$ bending v.
1330 – 1390	D-Band of CNT's
~1345	Adenine, Guanine activity
~1336	Adenine, phenylalanine CH deformation v.
1295 – 1304	C-O vibrations
~1302	CH$_2$ deformation modes
1238 – 1290	Amide III band
	CN stretching v., NH bending v., CO stretching, O=CN bending v.
~1267	Amide III (-helix, protein)
~1250	Amide III (-sheet, protein)
~1170	Weak CO-O-C stretching v.
1030 – 1060	C-O-P stretching vibrations, CO-O-C symmetric stretching v.; lipid related
~1056	RNA ribose C–O vibration
~1003	Phenylalanine (ring breathing)
930 – 960	RNA ribose C-O vibration
~965	CN asymmetric stretching v.

2.5.2.2 Signal Deconvolution

To analyze the spectral information, the spectra need to be de-convolved, revealing the actual contribution (a) of each peak at a specified location (x_p). The unique position of a convolved peak can be mathematically determined by double derivatization of the spectrum. A common method to de-convolve a spectrum is to fit the signal in certain areas with combined, weighted (g) Gaussian and Lorenzian functions described by Equation 2-20.

$$y = g*a*e^{-\left(\frac{x-x_p}{c}\right)^2} + (1-g)*\frac{a}{1+\frac{4*(x-x_p)^2}{c^2}}$$

Equation 2-20

This function is also described as a centred pseudo-Voigt function, approximating the Voigt profile of spectroscopic line shapes, generally found in spectroscopy, in which the spectral lines are broadened by a variety of effects, one of which alone would cause a Gaussian distribution of the otherwise Lorentzian or Cauchy distribution [148, 149].

2.5.2.3 Peak area ratio calculation

Peak area ratio calculations are commonly employed to give relative information between two, or a small number of peaks, in order to evaluate mixed or overlaid information quantitatively [150]. Ratioing the area of a peak versus another gives the advantage of the reduced influence of background variations, due to the self-normalization. It is a common method in pharmacology, chemistry and biology, [151] employing spectroscopic data for analysis [152]. It can be either applied to de-convolved or convolved spectroscopic data. The area of each peak of interest can be calculated by the formal integration of the deconvolution function (Equation 2-21) with the integration area in the limits between x_0 as the lower and x_e as the upper limit of the integral, or approximated by the sum of the intensities of the data points $y(x)$ in the selected area Equation 2-22. [130].

$$Ay_p = \int_{x_0}^{x_e} g*a*e^{-\left(\frac{x-x_p}{c}\right)^2} + (1-g)*\frac{a}{1+\frac{4*(x-x_p)^2}{c^2}} \, dx$$

Equation 2-21

$$Ay_p \approx \sum_{x_0}^{x_e} y(x)$$

Equation 2-22

2.5.3 Multivariate Data Analysis

2.5.3.1 Principal Component Analysis

Principal component analysis (PCA) is a method of multivariate analysis widely used with datasets of multiple dimensions. It allows the reduction of the number of variables in a multidimensional dataset, although it retains most of the variation within the dataset. To achieve this reduction, p variables ($x_1, x_2,..., x_p$) are taken and the combinations of these produce the principal components (PCs), $PC_1, PC_2,..., PC_p$, which are not correlated to each other and are sometimes called eigenvectors. This lack of correlation means that the PCs represent valuable different 'dimensions' in the data. The order of the PCs denote their importance to the dataset. PC_1 describes the highest amount of variation, PC_2 the second highest and so on. Therefore, var $(PC_1) \geq$ var $(PC_2) \geq$ var (PC_p), where var (PC_i) represents the variance of PC_i in the considered data set. Var (PC_i) is also called the eigenvalue of PC_i. With successful application of PCA, the number of PC's is reduced to a small number of eigenvectors or dimensions which describe the largest amount of variance in the dataset.. In a given dataset of p variables, which in this case are wavenumbers, and n samples, the squared covariance or correlation matrix can be calculated using the following equation (Equation 2-23).

Equation 2-23
$$Co(X_j, X_k) = \frac{\sum_{i=1}^{n}(X_{ij}-\bar{X}_j)(X_{ik}-\bar{X}_k)}{(n-1)} \quad \text{where } \bar{X}_j = \frac{\sum_{i=1}^{n} X_{ij}}{n} \quad \text{and } j,k=1,2,3,...,p.$$

The covariance matrix is applied when the variables of the dataset are on a comparable level. As soon as the variables of a dataset are in different units or have different scales, the correlation matrix is applied to standardized variables. PCA itself is the calculation of the eigenvalues and eigenvectors of the samples' correlation matrix, which can be calculated in an iterative process.

The first principal component (PC_1) is therefore a linear combination of the original variables $x_1, ..., x_p$, varying according to the individual feature, in this case the wavenumber, as much as possible, while the sum of the squared coefficients equals one. These coefficients $a_{11}, ..., a_{1p}$ are assigned to the original p variables of the PC_1 (Equation 2-24).

Equation 2-24
$$PC_1 = a_{11}X_1 + ... + a_{1p}X_p = \sum_{j=1}^{p} a_{1j}X_j \quad \cup \quad \sum_{j=1}^{p} a_j^2 = 1$$

Thus, the eigenvalue of PC_1 is as large as possible with the constraint of the constant a_{1j}, this constraint being necessary to avoid an increasing eigenvalue of PC_1 just by increasing one of the coefficient values a_{1j}. The actual number of PCs that feed into further analysis is dependent on the variance explained by the eigenvalue of a PC, visualized by a Scree plot, and the threshold one applies, in order to give a satisfactory representation of the original dataset, explaining most of the variance within it [120]. The slope of the declining eigenvalues of the first few PCs gives an additional indicator of the quality of the PCA. PCA results, after the mathematical transformation of the original data matrix, in the data matrix taking the form X=T.P+E, where T are the scores of the sample in the PCs and P are the loadings of the variance per variable over the whole dataset. E represents an error matrix which is usually not considered (Figure 2-19).

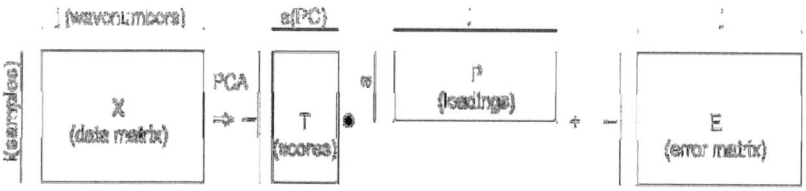

Figure 2-19 Schematic Illustration of PCA

Thus the scores of a sample in PCs are orthogonal to each other and are therefore uncorrelated. They represent coordinates along the dimensions of the PCs e.g. a three dimensional space for 3 PCs, used to access possible separation of certain groups within samples. Analysing the loadings of a PC can give information regarding the variables that are the source of variance in a given PC. Positive values of loadings are generally considered as indicating variables which contribute positively to the variance described by a given PC, while negative values contribute inverse to the variance described by the PC, and are therefore harder to explain. The PC's can be understood as abstract dimensions on which a sample can score, similar to the cartesian coordinate system though not being linear. Principal component analysis finds application in spectroscopy due to its ability to compress multivariate data into a number of variables of significance. It is capable of reducing the dimensionality of spectroscopic data which makes it easier represent them. PCA is a fundamental method in chemometric analysis used for example in the toxicological evaluation of pharmaceuticals [95] and monitoring processes during fermentation [153].

2.5.3.2 Partial Least Squares Regression

First described by Wold in 1960, partial least squares (PLS) regression is a popular and well known tool in the field of chemometrics [97, 154, 155]. The aim of PLS regression is the construction of a model to describe the response variables (i.e.

analyte concentration) in terms of the observed variables (spectra) from a set of training data. The least squares model is given by:

$$Y = XB + E \qquad \text{Equation 2-25}$$

where Y=n x m are the dependant variables (e.g. concentration), X=n x p are the independent variables (e.g. Raman spectra), B=p x m is the matrix of regression parameters for each component in Y and E is the matrix of residuals (differences between measured and predicted variables). PLS regression is similar to that of principal component analysis (PCA). PCA produces factors based on variance solely on the X matrix where the PLS regression algorithm considers both the X and Y matrices ensuring the factors correlate the X matrix to the (correlation) target. PLS regression differs from similar techniques such as multiple linear regression (MLR) and principal component regression (PCR) in the way that the X and Y variables are decomposed simultaneously maximizing the covariance between both matrices and allowing direct correlation between the spectra and a [155] classifier or descriptor for an expected behaviour of the sample. Commonly, these classifiers are termed targets. In addition to the scores and loading matrix, a series of weight vectors are calculated which enhance the variables with high correlation to the targets. The initial weight vector is calculated as follows:

$$w_1 = X^T y / \left(\left\| X^T y \right\| \right) \qquad \text{Equation 2-26}$$

The initial scores vector is calculated as:

$$t_1 = X w_1 \qquad \text{Equation 2-27}$$

and the loadings as:

$$p_1 = X^T t_1 / \left(\left\| t_1^T t_1 \right\| \right) \qquad \text{Equation 2-28}$$

The regression parameters are calculated as follows:

$$\hat{b} = y^T t_1 / \left(t_1^T t_1 \right) \qquad \text{Equation 2-29}$$

The residual matrix is calculated as:

$$E_1 = X - t_1 p_1^T \qquad \text{Equation 2-30}$$

The algorithm continues for each factor using E_1 instead of the weight matrix to calculate the second set of weights. When presented with an unknown spectrum, y is determined using W and P to compute scores for the unknown spectrum along with the regression parameters allowing the concentration of y to be determined from

Equation 2-25. PLS regression has been applied in a wide range of spectroscopic analyses, predicting various attributes of biological and non-biological materials [98, 122, 135, 156]. For example, this method has been used to predict the concentration of plasma glucose and lactic acid concentrations in the supernatant culture medium of human glioma cells [157]. It has also found application in the assessment of relative concentrations in mixtures of oral bacteria [158].

2.5.3.3 GA - Genetic Algorithm

Calibration models are known to be greatly improved through the application of efficient feature selection methods, increasing the predictive ability and reducing model complexity. One such method is the adaptive search technique known as the genetic algorithm (GA). In the first experiment, a GA based variable selection procedure is used to reduce the original spectra to a subset of wavenumbers to correlate Raman spectra to response. The first generation for evaluation is a random population consisting of a number of individuals or "chromosomes", each containing a subset of the original variables. Each chromosome is composed of a vector of 1s and 0s, corresponding to the wavenumbers in the X matrix, (1 if selected and 0 if not) where each wavenumber is termed a "gene". The performance of models resulting from each chromosome is determined by means of a fitness function (here the root mean square error of cross validation is used). Once each generation is evaluated, a new set of chromosomes is produced by retaining and "crossing" over the fittest individuals from the previous generation. "Mutations" are also produced which force the evaluation of new combinations avoiding saturation with similar sets of events and can further lower the number of variables and fitness values. The process continues until the difference in mean fitness levels between successive generations is below a certain threshold, whereupon the GA is terminated to avoid over-training and avoid over fitting risk in the PLS regression model [159-161]. Feature selection in this work was achieved using GA optimisation (with the genpls MATLAB toolbox by Ledardi) over 100 runs requiring approximately 60 minutes depending on the computational speed.

Table 2-5 Genetic Algorithm parameters

Parameter	Value
Chromosome size	30
Max. genes per chromosome	30
Mutation probability	0.01
Crossover probability	0.5
Preprocessing	None
Max LV	15
#runs	100

Each calibration model was evaluated using root mean squared error of cross validation (RMSECV) and root mean squared error of calibration (RMSEC) performed on the calibration set. The root mean squared error of prediction (RMSEP) of the independent testing set was also calculated (40% of the dataset).

2.5.3.4 HCA – Hierarchical Cluster Analysis

Hierarchical clustering groups data over a variety of scales by creating a cluster tree, usually displayed as dendrogram or tree [162, 163]. The tree is not a single set of clusters, but more likely a multilevel hierarchy, where clusters at one level are joined with clusters at the next level. This allows a decision of which level or scale of clustering is most appropriate. Hierarchical clustering builds (agglomerative), or breaks up (divisive), a hierarchy of different clusters. In the representation as a dendrogram (Figure 2-20) individual elements (a, b, c, d, e, f, leafs) are located at one end and a single cluster containing every single element at the other (a,b,c,d,e,f, …root). Agglomerative algorithms begin at the leaves of the tree, whereas divisive algorithms begin at the root. In Figure 2-20 (right) the arrows indicate an agglomerative clustering. Truncating the tree at a certain height will give a clustering at a selected precision, e. g. after the second row will give the clusters (a), (b c), (d e) and (f). Cutting behind the third row gives the clusters (a), (b c) and (d e f) resulting in a coarser clustering, with a reduced number of larger clusters. This method creates the hierarchy from the individual elements (leaves) by progressively merging clusters to the root. In this example six elements (a), (b), (c), (d), (e) and (f) are employed. The first step is to determine which elements to merge in a cluster. Usually, the closest elements are taken, according to the chosen distance. Optionally, one can also construct a distance matrix, where the number in the ith Row, j^{th}. column is the distance between the ith. and j^{th}. elements. Then, as clustering progresses, rows and columns are merged as the clusters are merged and the distances are updated. This is a common way to implement this type of clustering, and has the benefit of matching distances between clusters. In the application to data from PCA analysis, commonly the Euclidean distance between incidents (e.g. the scores of the PC's) along the dimensions of the principal components is employed for determination of the hierarchical conformation.

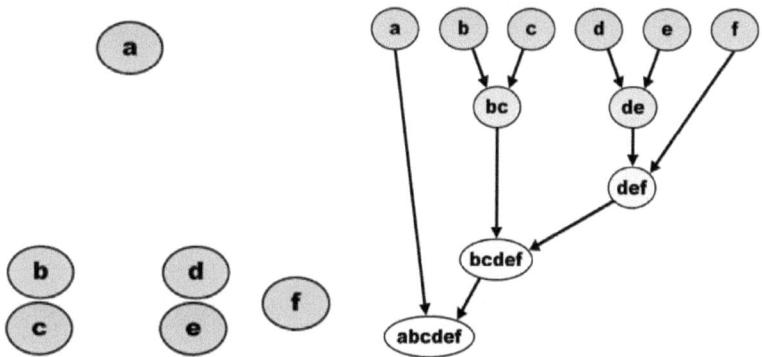

Figure 2-20 Scattered dataset (left) and dendrogram of a hierarchical clustering with Euclidean distance as distance metric (right) (http://en.wikipedia.org/wiki/Cluster_analysis)

2.5.3.5 Independent Component Analysis

Independent component analysis (ICA) is a powerful statistical tool. Its main field of application is on the one hand feature extraction and on the other hand blind source separation (BSS). In BSS, multiple linearly mixed signals are separated without knowledge about the mixing process, in order to extract the source signal. For example, two recordings of data measure the same incident from a different perspective x1(t), x2(t). These two recordings are equal to the weighted sums of the source signals s1(t), s2(t) and a set of linear equations is obtained Equation 2-31.

$$x_1(t) = a_{11}s_1 + a_{12}s_2$$
$$x_2(t) = a_{21}s_1 + a_{22}s_2$$
Equation 2-31

Normally, to solve the equation system to yield $s_1(t)$ and $s_2(t)$, the parameters $a_{11}...a_{22}$ must be known. These parameters depend on the difference in projection direction of the recordings. Thus one only has recorded signals $x_1(t)$ and $x_2(t)$ so one can only solve it with broad statistical assumptions. For independent component analysis it is sufficient to assume that both recordings are stochastically independent. Assuming x and s are independent of time but random variables, x_1 and x_2 become random samples. Additionally it can be assumed that x and s have the expectation value zero, which can be achieved by appropriate normalization. As in principal component analysis, the data set can be presented in the matrix form.

$$x = A \cdot s \qquad \text{Equation 2-32}$$

To calculate s with the knowledge of A it would be necessary to calculate $W=A^{-1}$. Without knowledge of A, W has to be calculated differently.

$$S = W \cdot x \qquad \text{Equation 2-33}$$

Because the components of x are usually highly correlated, it is possible to choose W from the principal component analysis in such a way that the correlation would have been lost. Thus, while principal component analysis uses the covariance matrix and second moment statistics, independent component analysis uses all moments. Therefore, independent component analysis not only removes correlations but it establishes the most possible statistical independence between components. Since s and A are unknown, any scalar value of the source s_j can be equalized by division of column a_j in the matrix A by this factor. Thus, as the variance E is not important for ICA, it usually is scaled E=1 [164]. Unfortunately, it is not possible to predict the order of the independent components. A and s are unknown and therefore the values in s can be swapped, leading to a new matrix A. Formally the permutation matrix P can be introduced. Together with its inverted form, Equation 2-33, changes to $x=A\ P^{-1}P.s$. The elements of P.s are the former components s_j in a different order. The matrix $A.P^{-1}$ is a new matrix generated by the ICA algorithm. The major constraint for application of ICA is that the Independent components must not be distributed normally. As previously mentioned, ICA has some similarities to PCA. In PCA the data are expressed in an orthogonal basis, whereas ICA expresses the data on a non-orthogonal basis. Therefore ICA is capable of capturing the data more efficiently. (Figure 2-21)

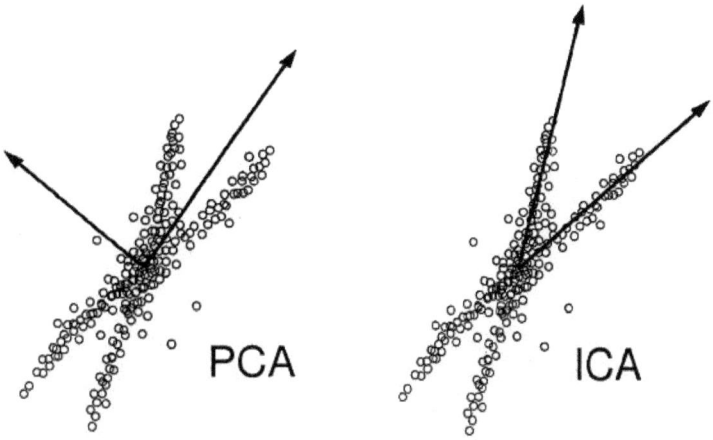

Figure 2-21 Comparison between PCA and ICA (Apo Hyvärinen, 2000)

Independent component analysis has been applied to spectroscopic data in the classification of tissue from human coronary arteries *in vitro* [165], and used for "de-waxing" tissue samples computationally [166]. For the present work, independent component analysis is expected to extract the individual toxic influences of each agent from the measured spectra in chapter 5. The extracted components may then aid the multivariate model of the overall toxic response, with reduced variance in partial least squares regression.

2.5.3.6 Hyper spectral Imaging

Spectral imaging is the simultaneous measurement of spatially encoded spectral information and the recording of the spatial location within a cell or tissue The images so generated are referred to as hyper-spectral or multi spectral images [167, 168]. They can be used to image and analyze individual cells and microscopically invisible nano structures e.g. single carbon nanotubes [169]. In Raman spectroscopy spectral measurements are recorded by rastering the laser focal spot over the sample as defined by its visual image. This creates a hyper cube of data that can be visualised and treated as a series of spectrally resolved images with each plane of the hypercube corresponding to a spatial image of the biochemical information at that wavenumber. . Alternatively, the selected plane may highlight a certain distribution of the sample compound given that the spectral signatures are a feature at the selected wavelength. Therefore, this imaging technology can be applied to univariate or multivariate analysis, delivering an optical impression comparable to microscopically visible image.

2.6 Summary

This chapter detailed the experimental aspects of the exposure of human epithelial lung cells to carbon nanotubes that will be assayed by Raman spectroscopy. It is emphasised that the study design is based on previous cytotoxicological studies, such that the application of Raman spectroscopy can be validated against gold standard techniques. The background theory and practical use of the equipment employed has been described and the mathematical methods for pre-processing of spectral data and univariate and multivariate analysis were introduced. In the forthcoming experimental sections of this work, in parallel with the experimental components, continuous progress was made towards optimisation of the signal preprocessing and analysis protocols using the techniques outlined in this chapter. Therefore each experimental section will detail which protocol was employed. Surprisingly, given certain constraints, e.g. the lack of comparability between spectrometers, and therefore the omission of SRM calibration, the necessity of noise filtering was reduced. Furthermore, by modelling the substrate in a noise free fashion, the requirements for noise reduction is further reduced. Not surprisingly, over the course of this work, it became evident that optimal results were obtained from signals of relatively high SNR_{db} which required minimum processing.

Chapter 3 : Exposure Study

3.1 Introduction

In this chapter, the effects of direct exposure of A549 cells to SWCNT dispersed in culture medium, monitored by Raman spectroscopy, are presented and analyzed. The study has been published as "Raman spectroscopy -- a potential platform for the rapid measurement of carbon nanotube-induced cytotoxicity" in the Analyst 2009 [170]. An initial attempt is made to correlate the acquired data with results from the clonogenic endpoints of a previous study [99] using univariate analysis. In order to corroborate the potential of Raman spectroscopy as a toxicity probe, correlations of toxic responses to spectral features identified in literature for exposure to mercury [107] are performed, based on fitted peak area ratios. Certain aspects of these results seem to fit well to the clonogenic endpoint of reduced proliferative capacity expressed by reduced colony size, as a function of exposure concentration. The results from multivariate analysis show a clear picture of dose dependent separation either in the pre processed plain data or doubly derivatized datasets.

3.2 Materials and Methods

3.2.1 Cell culture

Square Quartz slides (24.5mm x 24.5mm, UQG Optics Ltd.) were coated for 24 h at 4°C with a sterile solution of 2% gelatine (Type-B from bovine skin) in deionised water (dH_2O) solution. Such substrates have previously been shown to be optimal for cell growth and subsequent spectral analysis [91]. A549 cells were cultivated as previously described in 2.4.3 and harvested at 85% confluence. The reseeded cells were allowed to attach to prepared quartz substrates at a concentration of approximately 2×10^6 cells per slide for 24h. After the 24h incubation period, the unattached cells were rinsed off with PBS. The SWCNT were dispersed with an ultrasonic tip in four exposure suspensions (0 mg/l (control), 1.56mg/l, 6.25mg/l, 25.0 mg/l, 100mg/l) in supplemented medium. HiPco Carbon Nanotubes (Carbon Nanotubes Inc.) at these concentrations were employed for the study for consistency with previous studies [75, 99, 102, 171]. The ultrasonic tip was operated at a medium level of output for a total time of 30s in 10s sequential intervals to prevent sample heating [101]. The cells were then exposed to 3 ml of each of the different SWCNT suspensions for 96 hours. After the exposure period, the slides were rinsed with PBS and fixed in 4% formalin in PBS solution for 10 minutes, rinsed once again in dH_2O, and finally stored in dH_2O at 4°C prior to conducting the spectral measurements.

3.2.2 Spectroscopy

Raman Spectroscopy was carried out with the Instruments SA Labram 1B Raman confocal microscope using a 514.5 nm laser excitation with a grating of 1800 l/mm, providing a spectral dispersion of about ~1.65 cm^{-1}/pixel. Spectra were recorded using a water immersion lens (Olympus Lum-Plan FL 100x) from substrates immersed in water to prevent desiccation of the samples. The immersion reservoir was constructed by inserting a quartz window into the bottom of a Petri dish filled with dH$_2$O. The x100 water immersion objective produced a spot size of approximately 1μm in diameter at the sample.

All recordings were performed as an average of three individual measurements of one point to reduce the influence of spectral noise. The system was calibrated to the spectral line of crystalline silicon at a constant room temperature of 21°C. The measurement range was set to an interval of ~250-1750 cm^{-1} in order to detect spectra within the fingerprint region of the cell samples [91] and the characteristic SWCNT features [3]. Before spectral acquisition, the dark current of the system and the system intensity response were recorded. After recording a number of spectral measurements per slide, the spectrum of the substrate was acquired. The laser power was set to 23 mW at the sample and the acquisition time was set to 90 s. In total, ca. 75 spectra (25 per sample in triplicate) were recorded from the nuclear portion of multiple cells at each concentration.

Principal component analysis was employed to identify outlier spectra [172]. Cells across the whole area of the sample slide were chosen for measurement in an attempt to ensure a true representation of the sample in order to limit variability that might occur due to the spatial position of the laser focal spot within the nuclear portion of cells, and biological variability that could occur between samples of the cell line. It was noted, however, that even after repeated washing with PBS, some single wall carbon nanotube aggregates could be visibly observed attached to the cells, although no SWCNTs were observed inside the cells themselves [75]. All measurements reported here were taken away from regions where large aggregates were visible (Figure 3-1).

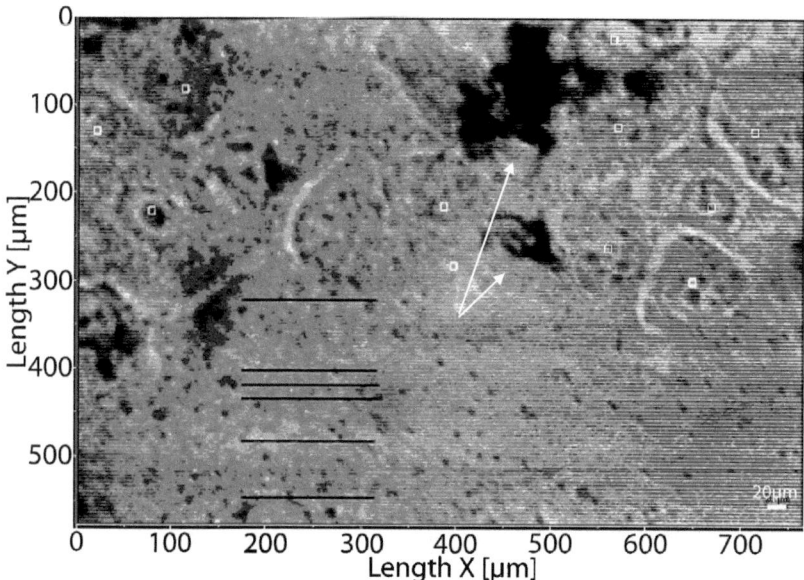

Figure 3-1 Micrograph of A549 exposed to 100 mg/l SWCNT with clearly visible aggregates (indicated by the arrows, the squares indicate the manually selected measurement acquisition points)

3.2.3 Data analysis

In total, 321 valid spectra were acquired for the 5 distinct concentrations (

Table 3-1) with a spectral range from 248 to 1751 cm^{-1}. The raw spectra were imported into Matlab 7.3 for pre-processing and analysis. Every spectrum was corrected for system intensity response, according to the guidelines of NIST (see Section 2.4.8.2) [173]. Prior to the subtraction of the underlying substrate (quartz) signature, each spectrum was scaled to the characteristic quartz feature at 486 cm^{-1} [144] and the background was subtracted manually. Finally, the spectra were cropped to a spectral window of 599-1700 cm^{-1} to isolate the fingerprint region. In order to minimise the noise the spectra were smoothed using the Savitzky Golay algorithm [124] with a 15 point window and a polynomial order of 3 prior to further analysis.

Table 3-1 Sample numbers after recording the measurements and outlier removal

Sample Concentration	Recorded Replicates			Validated Replicates		
	I	II	III	I	II	III
0.0 mg/l	25	25	25	22	21	20
1.56 mg/l	25	25	25	19	21	23
6.25 mg/l	25	25	25	22	20	21
25.0 mg/l	25	25	25	22	24	21
100.0 mg/l	25	25	25	21	23	21

3.3 Results

3.3.1 Univariate Analysis

The Raman spectrum of a SWCNT sample (suspended in water at concentrations similar to those used during this study) exhibits characteristic radial breathing modes (RBM) in the region of 100-300cm^{-1} [84] (Figure 3-2). These features describe the synchronous oscillation of the atoms of the nanotube in the radial direction and can be used to define structural characteristics of SWCNTs such as their diameter, metallicity, and helicity [148]. The so called "disorder-induced" D band appears at 1330-1390 cm^{-1} and is reputedly an indicator for disorder in the graphene sheet. The tangential mode, or G-Band, originating from tangential oscillations of the carbon atoms in the nanotubes, appears at 1583-1605cm^{-1} [85, 148, 174].

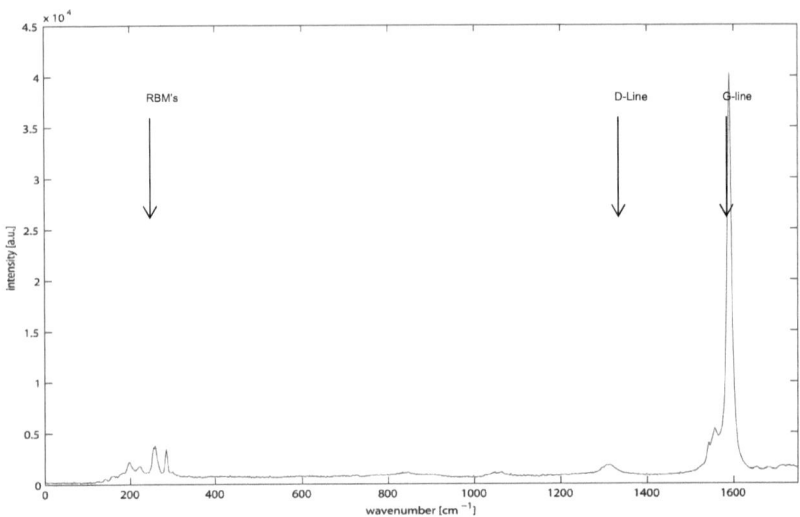

Figure 3-2 Raman Spectrum of SWCNT with characteristic features (RBM's at ~180-300 cm^{-1}, D-Line at ~1350 cm^{-1}, G-line at~1590 cm^{-1})

Spectra of A549 cells, (Figure 3-3 (b)) exhibit classic features associated with cellular material within the Amide I band area of 1656-1690 cm^{-1}, consisting of ~80% of CO stretching,, ~10% CN stretching and ~10% NH bending vibration modes, indicating protein based α-helix, random coil and β-sheet structures [105]. In the Amide III area at about 1238 cm^{-1}, β-sheet and random coil structures are indicated by ~30% CN stretching and ~30% NH bending vibrations, as well as ~10% CO stretching and ~10% O=C-N bending vibrations. Vibrational features of amino acids and amino acid hydro halides appear in the area of 1485 - 1660 cm^{-1} (NH deformation vibrations and α-form C=O stretching of polypeptides). Characteristic signals of lipids appear at 965 cm^{-1} (CN asymmetric stretching vibrations), 1170 cm^{-1} (weak CO-O-C symmetric stretching) and 1451 cm^{-1} (CH$_2$ scissoring and CH$_3$ bending vibrations) [93, 105, 106, 175]. As the samples were rinsed before fixing as described in section 2.4.7., it is assumed that all features are of cellular in origin.

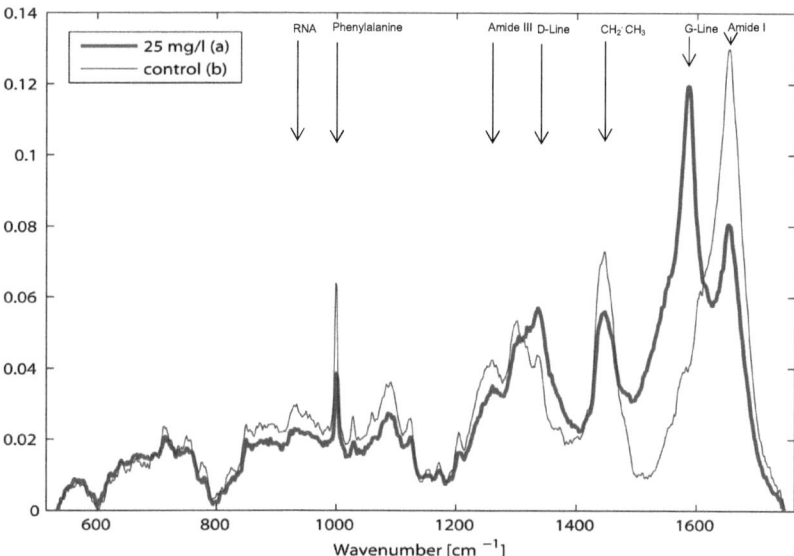

Figure 3-3 Raman Spectra of A549 exposed Cells (25 mg/l SWCNT) (a) and A549 control cells (b), filtered with Savitzky Golay Filter order 3, 15 points.

In Figure 3-3 (a), an average spectrum of a cellular sample exposed to a relatively high dose of SWCNTs (25mg/l) is shown. Strong contributions of the G-Line and D-Line features of SWCNTs as well as common cellular spectral features are clearly visible, although the SWCNT were washed off thoroughly and were not visible microscopically. In a previous study, in samples prepared under identical conditions, no SWCNTs could be observed internalized in the cells although small bundles or ropes were observed adhered to the cell surface [75]. The strongest peak of the typical SWCNT spectrum, the G-line at ~1585 cm^{-1} [148], overlaps strongly with the Amide I and water feature in this region of the cellular spectrum (1637,1656-1690 cm^{-1}) [105, 176]. This makes it difficult to utilize this band for analysis of cellular response to the SWCNT exposure without deconvolution. After background subtraction, the region of 1502-1700 cm^{-1}, was extracted and fitted with a series of mixed Gaussian/Lorentzian functions to extract the relative contributions of the SWNT G-line and the cellular Amide I band. Figure 3-4 (left) shows the intensity of the SWCNT G-line as a function of exposure dose in terms of concentration (mg/l). Although the Raman intensity is approximately linear as a function of dose up to

~30mg/l, the maximum dose of 100mg/l shows significant deviation from this. This apparent saturation of the response may be a result of over dosage, due to the nanotubes not being effectively dispersed throughout the sample, and/or an effect of the increased optical density of the residual carbon nanotubes which are resonant at the Raman wavelength, causing limited penetration of the light into the sample and also scattering of light by the sample. In Figure 3-4 (right) the intensity of the Amide I Raman band as a function of SWCNT dosage is shown. The intensity is seen to be only weakly dependent on dosage, indicating that the reduced intensity of the SWCNT G-line has origin primarily in saturation of dosage rather than optical effects, although the slight reduction at large doses points towards some optical effects. A direct visual comparison of the cellular Raman spectra demonstrates clearly that several individual peaks are altered as a result of exposure. Examination of the spectra reveals changes to the 1030-1060 cm^{-1} lipid bands caused by C-O-P stretching~ and C-O-O-C symmetric stretching ~vibrations [106, 175], an observation which correlates well with the work of Davoren et al. [75], which, using Transmission Electron Microscopy (TEM), demonstrated an increase of surfactant storing lamellar bodies in A549 cells after exposure to SWCNTs. This observation supports the assumption of a change in the overall lipid content in the exposed cells. Although the nucleus is primarily targeted, and the confocal volume of the laser spot has a depth of ~2-3 µm, some of the overlaying cytoplasm will be picked up. Changes to the RNA ribose C-O vibration at 930-960 cm^{-1} and 1295-1304 cm^{-1}, as well as changes to the adenine and guanine activity at ~1345 cm^{-1} [105, 176, 177] are also observed. It is possible, however, that the D-Line of the SWCNTs may be masking the response in this latter region, but changes to the cellular spectra are clearly visible after deconvolution of the Amide III band area, far from the spectral features of the SWCNTs, with its different conformations at approximately 1238 cm^{-1}, 1258 cm^{-1}, 1271 cm^{-1} by a combined Gaussian and Lorentzian fit, known as a pseudo-Voigt function [149, 178] to approximate the Voigt profile, with a total of nine individual centred peaks, given by the second derivative of the unfitted region [107].

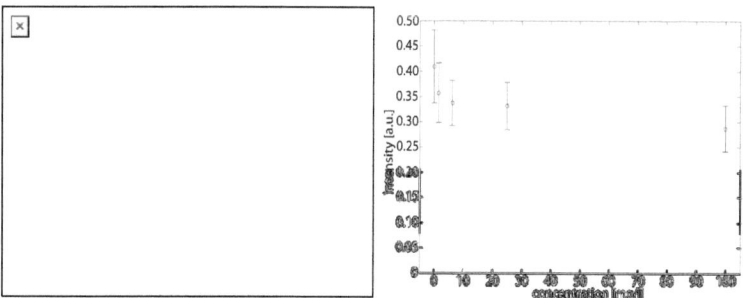

Figure 3-4 Intensity of G-Line at 1598cm^{-1} versus concentration (left), Intensity of Amide I at 1656cm^{-1} versus concentration (right)

Although it is difficult to precisely assign the many overlapping bands, the ratio of Raman peak heights for CH$_2$ deformation modes at ~1302cm^{-1}, DNA bases guanine, adenine and thymine at 1287 and 1338cm^{-1} versus the Amide III band at 1238cm^{-1} has previously been used for estimation of cellular toxicity [107]. Figure 3-5 shows the dose dependent response of these peak ratios. The three ratios exhibit a general trend of an overall increase with dose with the exception of the largest exposure dose of 100mg/ml which has been shown to have a saturated spectral/exposure response (Figure 3-5a). All spectral features, with the exception of the 1338cm^{-1} band are far from any SWNT bands, and the fact that the 1338cm^{-1} band exhibits the same trends indicates that there is minimal interference from the underlying tail of the SWNT D-line. Figure 3-5 D shows an approximately linear relationship between the ratio of 1338 cm^{-1}/Amide III as a function of G-line intensity which should more accurately represent the actual SWNT dose.

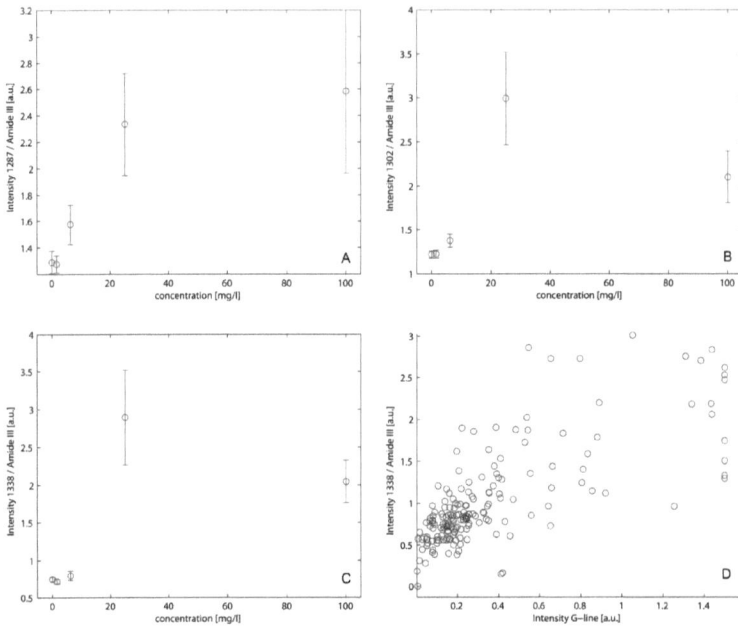

Figure 3-5 Peak ratios of 1287/Amide III (A), 1302/Amide III (B), 1338/Amide III (C),versus concentration and 1338/Amide III (D) versus the G-line intensity

The ratios of bands previously identified as cytotoxic markers clearly show a dose dependent response. This dose dependence correlates well with that previously observed for colony size in clonogenic assays on the same samples [77]. The dose dependent response of the colony size endpoint of the clonogenic study is plotted in Figure 3-6. A monotonic decrease in colony size with increasing dose up to ~100mg/l is observed. This toxic response has been attributed to a reduced proliferative capacity as a result of medium depletion caused by adsorption of components of the cell growth medium to the SWCNTs. Figure 3-6 demonstrates a clear correlation of the dose dependent ratio $1287cm^{-1}$ / Amide III with toxic response, as determined by the colony size endpoint of reference [77].

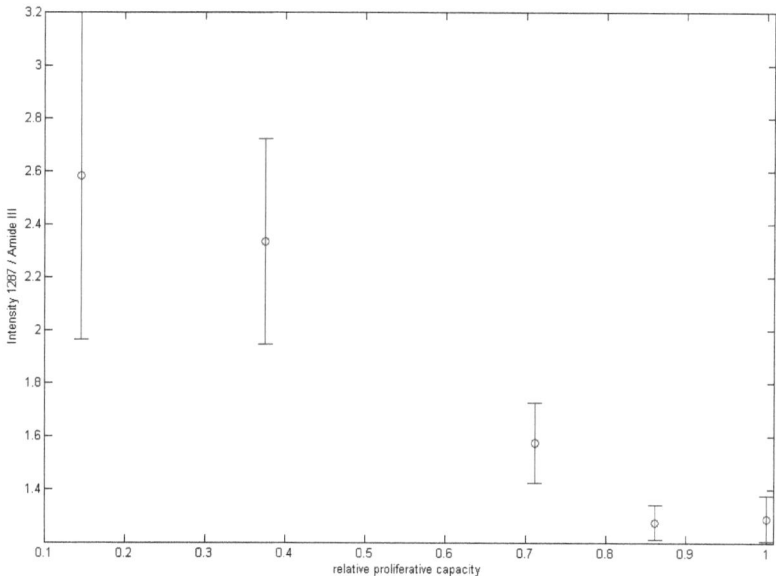

Figure 3-6 Correlation of the 1287cm^{-1}/Amide III peak ratio with colony size endpoint.

The results clearly indicate that dose dependent spectral markers can be identified in the Raman spectra of cellular samples exposed to SWCNTs. However, the intrinsic influences of inhomogeneity of the spatial dispersion of SWCNTs in e.g. cell culture medium [102] and the SWCNT residues adhering to the cells, as well as the complex changes to the spectral response of the cells, demand more elaborate data analysis methods, moving from the univariate approaches described above to the analysis of the spectral data by multivariate analysis. Principal component analysis will thus be employed as a more powerful classification tool, potentially elucidating a more detailed signature of the cellular response.

3.3.2 Multivariate Analysis

The loadings from the PCA of the spectral data (Figure 3-7) are used to monitor the spectral features according to their contribution to the variance in the dataset. The highest variance, describing 68.2 % of the overall variance, is seen in PC1 which is dominated by the strong features of the SWCNTs. The next largest variances seen in PC2 and PC3 are postulated to be related to biological responses associated with exposure to CNTs, although they only represent a further 25% of the total variance.

Within the first five components, component three has positive loadings associated with some the main features of the control spectrum (see Figure 3-3a), and those positive loadings seen at ~1030, ~1300,~1450 cm^{-1} are all associated with lipid vibrations, corroborating the peak ratio analysis of Figure 3-5. Also, loadings in the region 1230cm^{-1} to 1350cm^{-1} which are associated with the Amide III band, those at 1287 cm^{-1} and 1338cm^{-1} associated with DNA bases guanine, adenine and thymine, and lipid deformation modes at 1302cm^{-1}, feature strongly positive in PC3 and PC5.

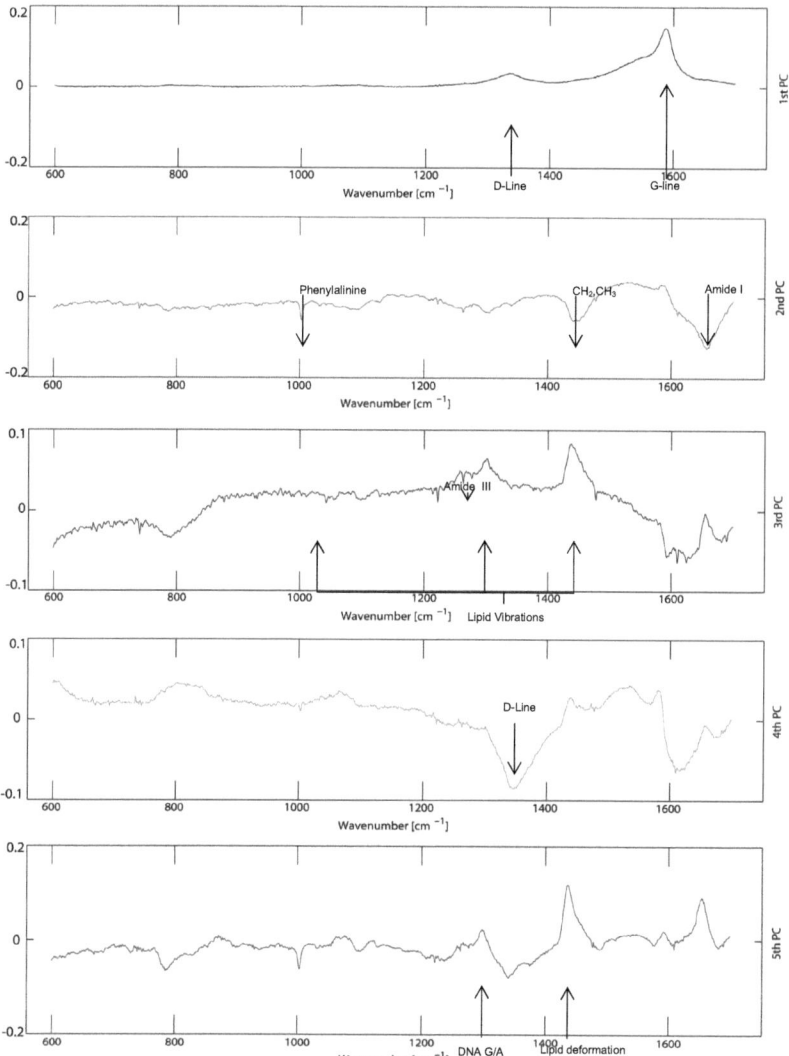

Figure 3-7 Individual principal component loadings plot of the first 5 components, (PC$_1$-PC$_5$ with explained variance of 68.2%, 20.3%, 4.5%, 2.5%, 1.6%)

The loading of PC$_2$ is very similar to an inverted cellular spectrum. By understanding the PC's as dimensions along which the samples score, the negative loading of PC$_2$ (as it is the dimension along which the exposed and unexposed samples separate) explains how different the exposed samples to average control samples are. The

PCA scores plot, Figure 3-8, shows a degree of separation into two classes between exposed and unexposed populations where the separation is caused basically by scoring positive or negative along PC_2.

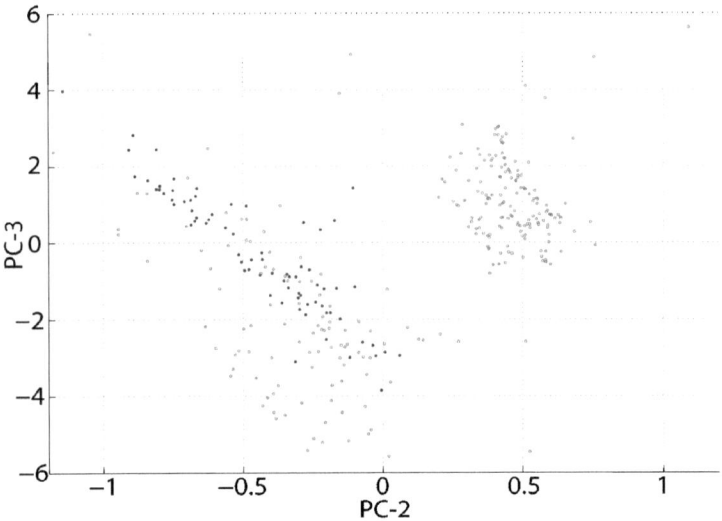

Figure 3-8 Principal component score plot of PC_2-PC_3 for every exposed (red) and control population (blue)

It is clear from visual observation at high concentrations [75] and the variability of the contribution of the G-line as shown in Figure 3-5 (D), that the spatial distribution and thus local concentration of the SWCNTs varies considerably from point to point in the sample at each dose. The separation or distinction between the five different exposure doses is not therefore very clear and a continuous variation of dosages as measured using the high spatial resolution of the laser is inferred. A clear distinction between exposed and non-exposed is however evident. By doubly derivatizing the data, the scores plot of the PCA shows distinct separations down to the level of exposure concentration of the samples, giving a defined cellular response and spatially denser co-localization of each group Figure 3-9. The plot demonstrates a well defined dose dependent response but again highlights the difficulties of establishing completely homogeneous exposure doses.

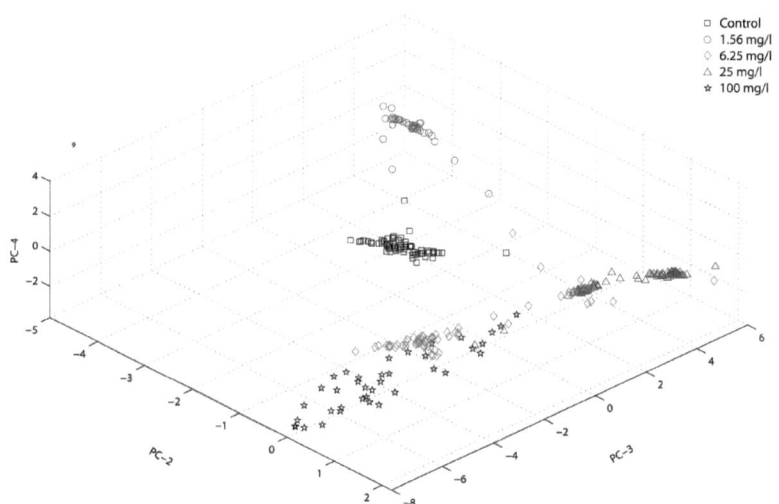

Figure 3-9 Principal Component Score Plot of PC_2-PC_4 for every spectrum of exposed concentration (control, 1.56, 6.25, 25.0, 100.0 mg/l) after being doubly derivatized

For illustration purposes, a PLS regression model was constructed. The training data for construction of the PLS regression model was formed by 60% of the full spectral dataset. The remaining 40% formed the test data set to evaluate the model. A GA was applied to reduce the number of wavenumbers required for prediction. The aim of the genetic algorithm was to minimize the RMSECV for the calibration model in predicting the clonogenic endpoints of CNT induced toxicity. The GA was performed over 100 runs. The fittest individuals used 178 variables, reducing the original dataset by 559 wavenumbers. In order to choose the optimum number of LV's to be retained 'leave-one-out' cross validation was carried out on the calibration set. (Figure 3-11) shows the results of the cross validation. Ten LV's were retained for model construction as the RMSECV did not decrease significantly after this point.

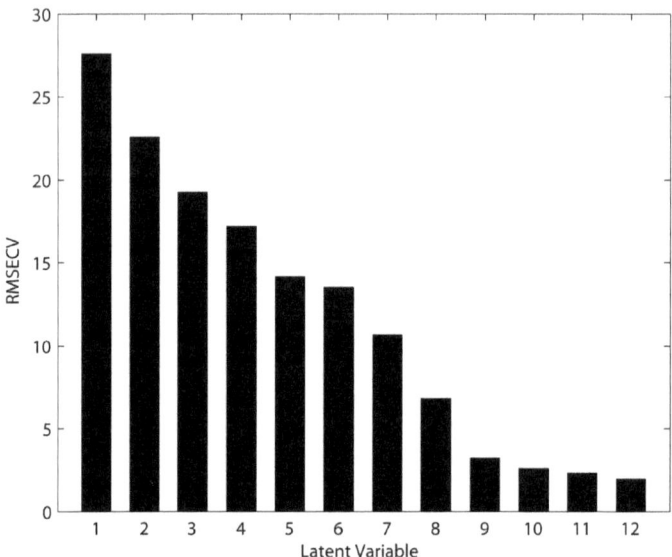

Figure 3-10 Cross validation results, the lowest RMSECV was observed at 10 latent variables (RMSECV = 2.53).

Table 3-2 shows the performance of the GA PLS regression.

Table 3-2 Performance of GA optimised PLS regression.

	#wavenumbers	LVs retained	RMSECV	RMSEC	RMSEP
PLS regression	737	10	4.31	3.17	3.37 (R^2 = 0.99)
GA-PLS regression	149	10	2.53	2.10	2.78 (R^2 = 0.99)

Using 10 latent variables, the GA PLS regression clearly outperforms multivariate calibration using the entire wavelength range, showing a decrease in all RMSE values. The independent testing set held back from training was used to determine if over fitting had occurred. Upon presentation of the testing set, the RMSEP was calculated to be 2.78 indicating an accurate model, and furthermore no over fitting was observed. Therefore, an accurate GA optimised PLS regression model has been created correlating Raman spectra to clonogenic endpoints (Figure 3-11) thereby reducing toxicity analysis time and the cost of analysis by negating the need for post exposure cell culture. While the RMSE values observed in this study are encouraging, a further reduction in these values would be beneficial.

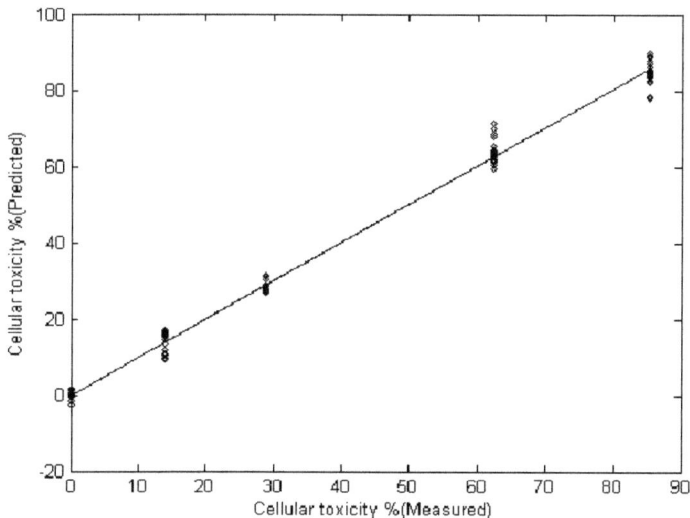

Figure 3-11 GA optimised PLS regression model correlating Raman spectra to clonogenic endpoints

3.3.3 Results from hyper spectral Imaging

In a trial study, spectral imaging of spectroscopic measurements was performed over an area of the samples where bundles of carbon nanotubes were, despite thorough washing, clearly visible (Figure 3-12). Over an area of 1720 µm², 1936 spectra were taken. Each spectrum consisted of the spectral range of 1500 cm^{-1} (200-1750cm^{-1}). In comparison to the visual image, two main features of CNT's are displayed, colour coded, representing their spectral intensity at certain planes. These spectral 'slices' of the hypercube at ~1331cm^{-1} (D-Line) and ~1590cm^{-1} (G-Line) display the distribution of CNT's within areas which were not visually conspicuous (Figure 3-13). It is important to highlight that the acquired data (2D) were recorded over a significantly reduced measurement time of 10s and much lower laser energy (3mW) at the sample to avoid saturation of the CCD detector of the spectrometer due to the strength of the CNT's signal.

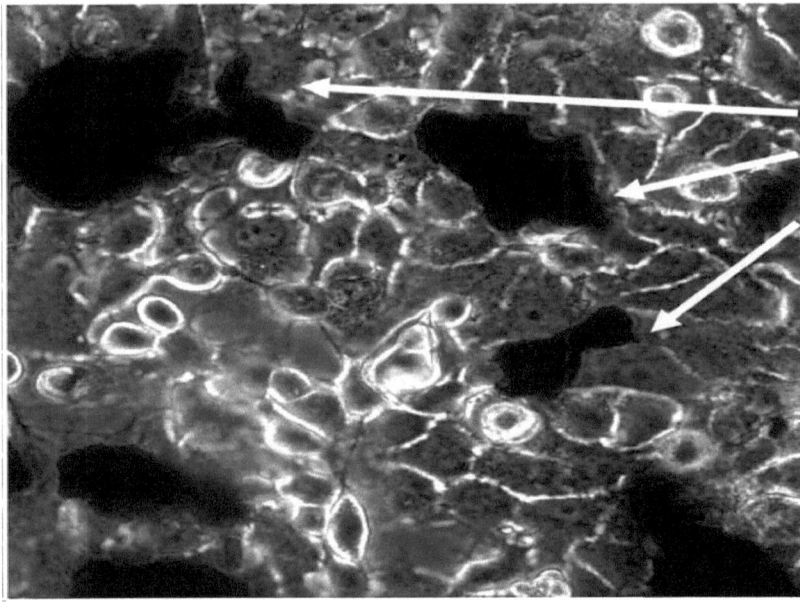

Figure 3-12 Phase contrast micrograph (200x) of A549 cells following 24 h exposure to 800 µg/ml SWCNT with 5% serum showing aggregates (arrows) on cell surface. These aggregates were still observed on the cell surface after several washes with PBS [Davoren et al. 2006]

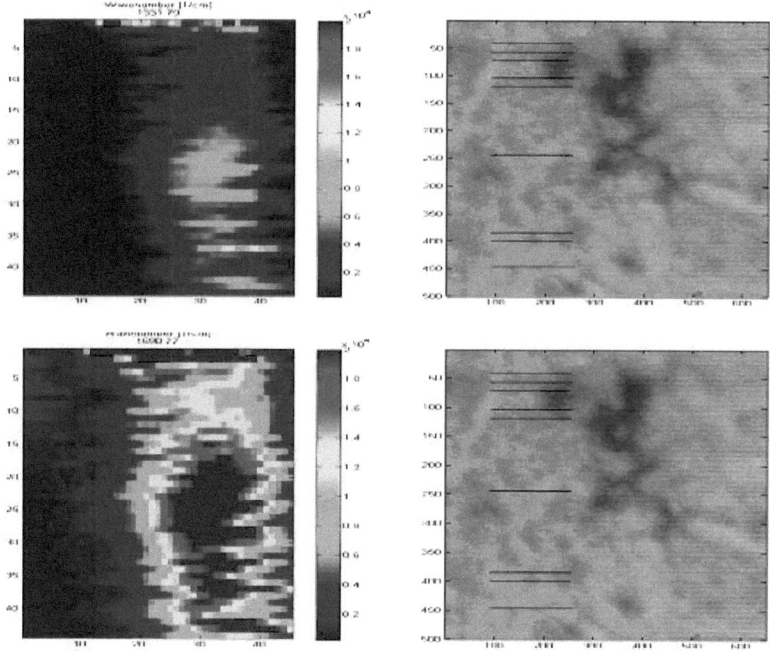

Figure 3-13 Spatial distribution map of CNT influence on A-549 as measured by Raman spectroscopy at indicated by the D-line feature (1331.79cm^{-1}) top row and the G-Line feature (1590.77 cm^{-1})bottom row

3.4 Conclusions

The potential of Raman spectroscopy as a viable tool to assess toxicology is demonstrated. Raman spectroscopy can be considered as a suitable technique for monitoring CNT induced biochemical changes at the cellular level. A good correlation is seen between markers of toxicity identified by a previous study and the exposure dose. A more refined control of cellular exposure coupled with other multivariate analytical techniques [178, 179] could extend this study to a truly quantitative assessment of the toxic response. The present study uses the same cell populations described in Herzog et al. [77] and the cells under investigation here were identical in exposure to those used in the previous work. A similarly good correlation between the spectral markers and the clonogenic endpoint of proliferative capacity is observed, indicating that the technique can overcome the previously identified problems with colorimetric assays in determining the cytotoxicity of carbon nanotubes. The changes in the spectra are visibly observable, dose dependent, and throughout are well related to cytological data, emphasizing that Raman spectroscopy is a potential analytical method for the examination of chemical and biological properties of cells. Principal Component Analysis as such shows good dose dependent separation of spectra. The combination seems to deliver a promising starting point for establishing a new kind of toxicological analysis, with the overall objective being the understanding and prediction of biological responses non-invasively with vibrational spectroscopy.

The study demonstrates the capabilities of Raman spectroscopy to detect cell alterations to A549 lung that were directly exposed to single wall carbon nanotubes. It was shown that key features of those CNTs can dominate spectroscopic results, but still, biological responses were notable. The difficulty of single wall carbon nanotube dispersion was highlighted. In a multivariate approach, the clear dose dependent response was utilised to build a predictive model for the exposure. A correlation could be established between spectroscopic markers and the overall toxicity of single wall carbon nanotubes. Though no single marker could be found and the response appeared to be more than one dimensional, the toxicity of SWCNT was classified in primary, direct, and secondary, indirect, toxicity. It was pointed out that for a multivariate modelling the application of a genetic algorithm for feature selection is beneficial. Furthermore the high sensitivity of Raman spectroscopy was demonstrated.

Chapter 4 : Secondary toxic responses as a result of medium depletion

4.1 Introduction

Previous studies have indicated that the toxic response of cells to the *in vitro* exposure to SWCNT is not only induced by primary, direct toxic influences, but that secondary or indirect toxic effects are apparent [102]. One of these, medium depletion, is a result of the adsorption of components of the cell culture medium by the SWCNTs, resulting in a cell growth environment which is deficient in essential nutrients. In the study by Casey et al. [101], colorimetric and clonogenic assays were employed to determine the influence of this depletion on cell viability. The depletion was created by the preparation of real exposure concentrations, which were then centrifuged and filtered to remove the SWCNTs and the adsorbed components. The cell cultures were exposed to these filtered solutions in order to mimic the concentration dependent medium depletion while eliminating any primarily toxicity due to direct exposure of the cells to the SWCNTs. Both colorimetric and clonogenic assays indicated that medium depletion effectively resulted in starvation of the cell cultures, compared to controls, resulting in a reduced cellular viability and proliferative capacity [2]. As colorimetric assays have proven to be problematic in the analysis of SWCNT toxicity [171], and clonogenic assays do not provide any direct evidence of underlying cellular mechanisms, Raman spectroscopy seems a promising alternative as a toxicological probe [170].

In this chapter, an identical exposure protocol to the experiments of Casey et al., is employed to explore the capabilities of Raman spectroscopy as a probe for such indirect toxic responses. Mirroring the comparison study of the previous chapter, univariate and multivariate analyses are again employed to correlate the spectroscopic data with the given colorimetric endpoints derived from the original cytotoxicological study. Significant improvement in the quality and reliability of the spectral data is achieved by evolving the methods of pre and post processing as described in chapter 2, in comparison to the study described in chapter 3. Although the resulting correlation of spectral data with colorimetric endpoints is poor, due to identifiable difficulties in the exposure protocol, indicators for the depletion dependent cell death and proliferation can be identified and the sensitivity of Raman spectroscopy is emphasized.

4.2 Materials and Methods

4.2.1 Sample preparation

A549 cells were cultivated as described in section 2.4.3 and harvested at ~85% confluence. As substrates for spectroscopic measurement, polished uncoated quartz slides (Crystran Ltd. Poole, UK) were used to remove any possible variance due to changes in coating thickness. The substrates were washed with ethanol, air dried under laminar flow, then loaded into six well plates (Nunc A/S, USA), and then were covered with 3 mL cell culture medium as described in section 2.4.5. The prepared chambers then received ~2.5 x10^4 cells each in the center of the quartz substrate and the cultures were incubated for 48 hours to allow the cells to attach to the substrates. Afterwards, the medium was removed and the samples were washed with PBS three times before exposure to the toxicants. Single walled carbon nanotubes were dispersed in cell culture medium with an ultrasonic tip for 30 seconds in a 10-second on-off interval, preventing heating of the SWCNT solutions, for the preparation of the stock solutions as described in section 3.1.2.1. HiPco Carbon Nanotubes (Carbon Nanotubes Inc.) were employed for the study for consistency with previous studies [75, 99, 102, 171]. From an initial concentration of 0.8 mg/ml, the exposure suspensions were serially diluted into seven different concentrations (0.00156, 0.00312, 0.0625, 0.25, 0.1, 0.4, 0.8) mg/ml. After preparing the SWCNT suspensions, they were stored for 24 hours at 4°C. Prior to application, these suspensions were centrifuged at 3000 rpm/1800 G for 20 minutes. Finally, to remove the dispersed nano particles, the SWCNT suspensions and cell culture medium were filtered using sterile cellulose acetate filters with a pore diameter of 0.2µm (Anachem ALG422A). This protocol is replicated from Casey et al. [101] and has been demonstrated to produce a medium in which the SWCNTs have been effectively removed, and furthermore has been depleted of much of its molecular components, which have adsorbed to the SWCNTs. For comparison, a sample of control, unexposed medium, was centrifuged and filtered using the same protocol. A further control of unfiltered medium was employed. The cells were then exposed to these filtered, depleted, media for 96 hours at 37.5°C (5% CO_2). After the exposure period, the slides were rinsed with PBS and fixed in 4% formalin in PBS solution for 10 minutes, rinsed three times with deionised water and finally stored immersed in dH_2O at 4°C until spectroscopic measurement.

4.2.2 Spectroscopy

Raman Spectroscopy was carried out with the Horiba Jobin Yvon Labram HR800 Raman confocal microscope, using the second harmonic (532 nm) of a Nd^{3+}:YAG laser as source, with a grating of 300 l/mm, providing a spectral dispersion of approximately 1.43 cm^{-1} per pixel. Cellular spectra were recorded using a water immersion lens (Olympus Lum-Plan FL x100) from substrates immersed in water in a sealable Petri dish with an opening for the lens to prevent evaporation of deionised water. The x100 water immersion objective produced a spot size of approximately 1μm in diameter at the sample. The system was calibrated to the spectral line of crystalline silicon at a constant room temperature of 21°C. The measurement range was set to an interval of ~245-1755 cm^{-1} in order to produce spectra within the fingerprint region of the cell samples [91]. Each measurement was integrated over 90 seconds using a laser power of 37 mW at the sample. Initially, 2 mL of each depletion medium concentration was measured in a quartz cuvette, to ensure that the filter strategy successfully removed all carbon nanotubes from the depleted medium. The substrate spectrum per slide was measured before and after acquiring the cellular spectra, in order to provide an indicator of any possible spectral drift during the measurement. The locations of 11 acquisition positions per cell were evenly spread along a straight line across the nucleus and adjacent cytoplasm. All recordings were performed as an average of three individual measurements of each point to reduce the influence of spectral noise. Four cells per sample were mapped in this way and the process was repeated in triplicate.

4.2.3 Data pre processing

The preprocessing protocol employed comprised of the following steps. Firstly, data were loaded into Matlab 7.3. The background of all spectra was removed with the rubberband function as described in section 2.4.8.6.3. Afterwards, all spectra were linearised as described in section 2.4.8.3. All spectra were then scaled to the characteristic quartz feature at 486 cm^{-1} [144]. Any potential drift between spectral measurements was analysed by cross correlating the pre- and post- recorded substrate spectra per slide. In the cases where drift was apparent, it was linearly interpolated over the time between the measurements and then integrated into the sample spectra by shifting the abscissae accordingly. The substrate spectra then were deconvoluted using a number of combined Gaussian and Lorentzian functions in order to model the substrate as described in section 2.4.8.7.3. The contribution of this model substrate was then calculated for each spectrum and subtracted. As all measurements were recorded in water immersion and the contribution of water inside the confocal volume might be variable, it was decided to deconvolute the Amide I region at ~1656 cm^{-1} with a series of combined Gaussian and Lorentzian functions to remove the water contribution at about 1640 cm^{-1}, as described in section 2.4.8.7.4. Finally, the spectral window was cropped to 600 - 1744 cm^{-1}, to remove any substrate residues caused by the variance of the quartz feature and the influence of the varying Rayleigh scattering. Outlier detection was performed manually after the preprocessing of the spectra by analysing each mean spectrum per cell and 957 spectra were deemed valid for analysis (Figure 4-2).

Table 4-1 Sample numbers (depletion study) after recording the measurements and outlier removal

Sample Concentration	Recorded Replicates			Validated Replicates		
	I	II	III	I	II	III
0.0 mg/l (control)	44	44	44	44	33	33
0.0 mg/l (filtered control)	44	44	44	44	44	33
1.56 mg/l	44	44	44	33	33	44
3.12 mg/l	44	44	44	33	33	33
6.25 mg/l	44	44	44	33	33	44
25.0 mg/l	44	44	44	33	33	33
100.0 mg/l	44	44	44	33	33	33
400.0 mg/l	44	44	44	33	33	33
800.0 mg/l	44	44	44	44	33	33

4.3 Results

4.3.1 Influence on the medium

The influence of the carbon nanotubes on the exposure medium was visually apparent after filtration. The higher the concentration of carbon nanotubes before filtration, the more colourless the filtered medium (conc > 25mg/L) becomes (Figure 4-1). The higher concentrations of 400 mg/L and 800 mg/L resulted in a completely (visually) colourless medium. The characteristic pink colour is caused by the pH indicator, phenol red, within the medium. However, the pH level is not changed by filtration as already reported [101]. This is one of the first indicators of the depletion of the medium by single wall carbon nanotubes. In addition, Raman spectroscopic analysis confirmed that no measurable amount of carbon nanotubes, or their aggregates, had passed the filtering process. The spectra of the depleted medium do not show any evidence of the characteristic sharp G-line feature of SWCNT's (Figure 4-2). This G-Band, or tangential mode, usually appears at 1583-1605cm^{-1} (Figure 3-2) and originates from tangential oscillations of the carbon atoms in the nanotubes [85, 148, 174]. Remarkably, besides the actual spectral content, the spectra looked very smooth even without any pre-processing and therefore no smoothing was applied.

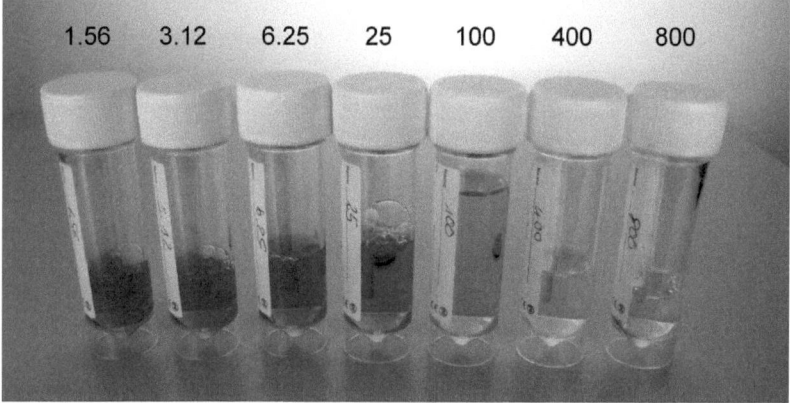

Figure 4-1 Photo of depleted media samples at a range of concentrations (notation in mg/L)

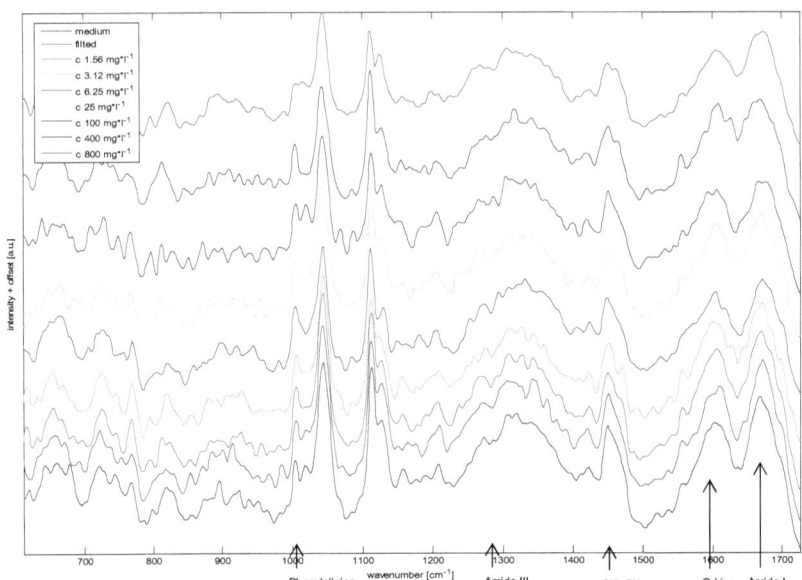

Figure 4-2 Raman spectra of filtered medium depleted by a series of SWCNT concentrations.

The Amide I band at 1656 cm^{-1} is a common indicator of protein in biological spectra as described in section 2.4.9.1. In this spectral region, among others, the spectra of the exposure media display a clear trend of reduced intensity with increased SWNT concentration (Figure 4-2), indicating protein adhesion to the SWCNT, and subsequent removal by the filtering protocol. The reduction in protein content is not a result of the filtering process alone, as the filtered control has Amide I features of similar strength to the control (Figure 4-3). Thus, as the SWCNT were not detectable spectroscopically, it is assumed that the SWCNT adsorbed a nutrient corona similar to other nanoparticles [180] and these and other aggregates are accumulated in the discarded filters. Remarkably, at high concentrations above 100mg/L, it seems that no further decline in signal intensity takes place. This observation is in good agreement with the results of Casey et al. [101], which indicate that significant changes in cell viability are notable only at very high concentrations. However, it is also noted that the total reduction of protein content of the medium, even at high SWNT loading concentrations, is ~30% and so the depletion is only fractional.

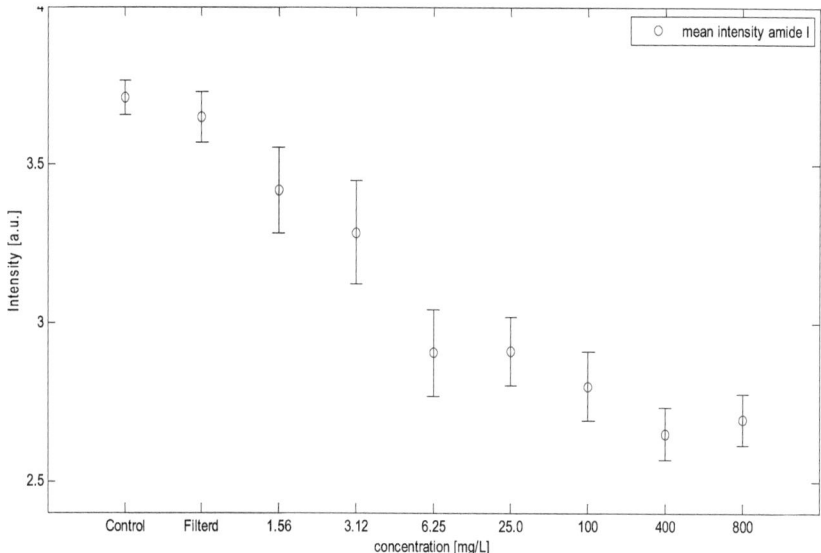

Figure 4-3 Reduced Amide Intensity with increased concentration of SWCNT (n=3 per concentration)

4.3.2 Impact of medium depletion on the cells

4.3.2.1 Univariate Analysis

Figure 4-4 shows mean spectra per line map of cells grown in unfiltered medium (control), filtered medium (filtered control) and the filtered media previously loaded with varying concentrations of SWCNTs. Notably, no features of single walled carbon nanotubes are detectable, further confirming that only indirect exposure to SWCNT took place.

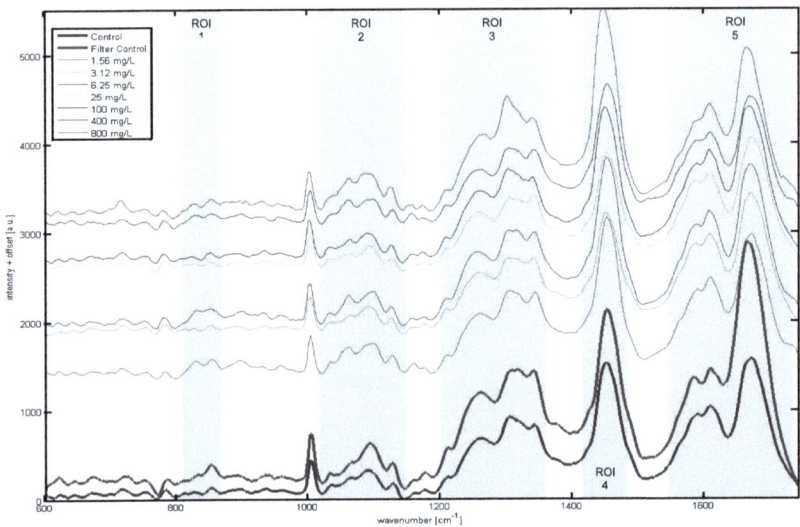

Figure 4-4 Mean spectra of A549 lung cells exposed to medium depleted by different concentrations of SWCNT (n > 33)

Although there are no striking differences between the mean spectra in terms of the dominant characteristic peaks, some smaller features do seem to be altered. For example, the feature at about 830 cm^{-1}, observable in the SWNT exposed media but not in the control or filtered control, suggests a slightly altered intensity of the O-P-O backbone and symmetric stretching vibration of DNA and RNA. The neighbouring feature at about 853 cm^{-1}, characteristic of the C-C stretching vibration of the amino acids proline and tyrosine, becomes more prominent in comparison to the control and filter control. Furthermore, the peak ratios in the cluster between 1019 cm^{-1} – 1145 cm^{-1} change with depletion, in particular the relative intensities of the C-O stretching vibration of RNA ribose at about 1056 cm^{-1} and the adjacent C-C chain vibration at about 1065 cm^{-1}. Whereas this cluster has only a small feature at about 1060 cm^{-1} in the control and filter control samples, this feature becomes notable in SWCNT depleted medium exposed samples. Corroborated by previous results which attributed a change in this spectral region to an alteration in the lipid contribution to the cellular composition [106, 175], this spectral region seems to be promising for further analysis as a possible co indicator for increased vesicularisation of the cell membrane, a lipid double layer, prior to cell death. The next cluster of interest is the spectral region between 1200 cm^{-1} and 1400 cm^{-1}, containing the Amide III band between ~1238 cm^{-1} and ~1284 cm^{-1}, which has in previous studies already functioned as a marker for toxicity in direct exposure to toxins, and which was employed in the study of the direct toxic effects of SWCNTs in the previous chapter [107]. It contains a strong CH$_2$ twisting vibration attributed to DNA, RNA and lipids at about 1302 cm^{-1} and a stretching vibration of CH at about 1342 cm^{-1} attributed to proteins, DNA and RNA. The next features of interest are the CH$_2$ deformation vibration attributed to lipids and proteins at about 1450 cm^{-1} [93, 105, 106, 175] and a DNA attributed feature at about 1580 cm^{-1}, the C=C bending vibrations of proteins at about 1610 cm^{-1}, and the Amide I peak at about 1656 cm^{-1}. The last region of interest is the very small feature at about 1743cm^{-1} at the extreme of the recorded spectral window, representing the C=O vibration of ester bondings in lipids, sometimes employed as an apoptotic marker in connection with the width of the Amide I band [181]. Overall, the filter control cells look very similar to the controls, probably indicating no or little change to the medium by filtration.

Despite the visually apparent variations in the spectral profiles of Figure 4-4, the markers for toxicity employed in chapter 3 (the ratio of 1287 cm^{-1}, 1302 cm^{-1} and 1338 cm^{-1} to the Amide III band at 1238cm^{-1}) indicate that no systematic dose dependent toxic response has taken place (Figure 4-5). The standard deviation of each point is however larger than the expected range of variation.

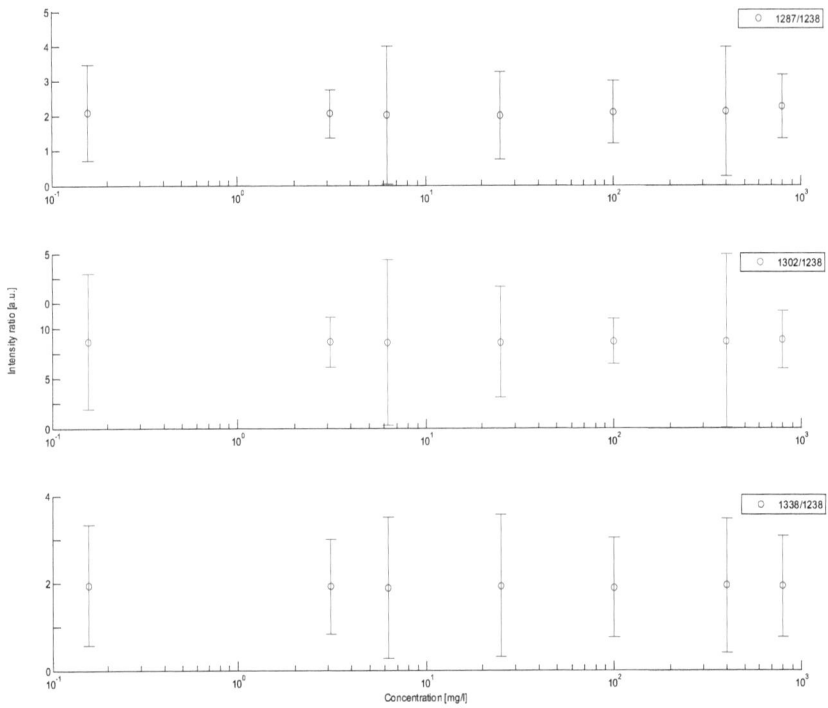

Figure 4-5 Peak ratios of 1287/Amide III (A), 1302/Amide III (B), 1338/Amide III (C), versus concentration

All filtered spectra were then normalised to the CH_2 feature at ~1450 cm^{-1}, a common means in spectroscopy to compare spectra [93]. This feature may be attributed to bending and scissoring vibrations in proteins and phospholipids [105, 182], and as CH and CH_2 are the most common groups in biological compounds [183], the feature can be used to normalize the spectra for their overall biological content [93]. The CH_2 feature at ~1450 cm^{-1} itself seems not to vary as a function of medium exposure (Figure 4-6), although, again, the standard deviation of the intensity measurement at each dose is very large. No obvious dose dependence, (filter control set to 100%) of identified cytotoxicity markers within the normalised spectra, was apparent, however, although the intensity of the Amide III band at 1238 cm^{-1} appears to be somewhat systematically and significantly reduced with increasing concentration. Similarly, the feature at 1234 cm^{-1} seems reduced significantly at the highest exposure concentration (Figure 4-7). The strong variation of the standard deviation per ratio can be explained by the partially covariant nature of the features with the intensity of the CH_2 feature.

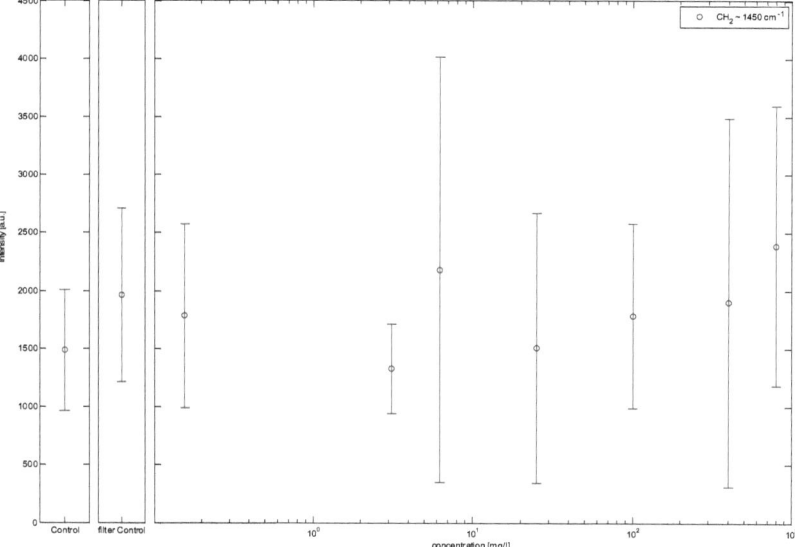

Figure 4-6 Variation of the CH_2 feature at ~1450 cm^{-1} versus depletion concentration as a measure for overall lipid and protein content (error bars indicate the SD per datapoint).

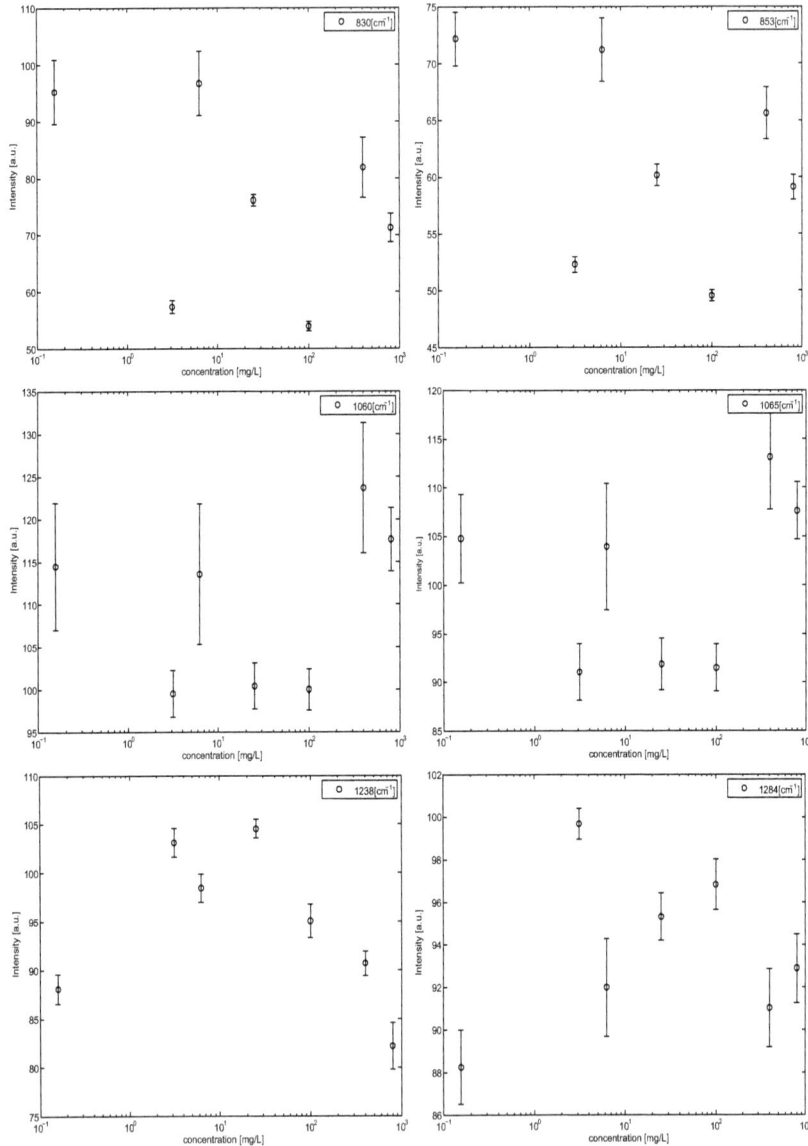

Figure 4-7 Extracts (I) of peak intensity ratios versus CH_2 (1450 cm^{-1}) with the filter control (0 mg/L SWCNT depletion) set to 100 %.

Remarkably, in comparison to the intensity of the filter control (100%), the feature at 853 cm^{-1} remains significantly below 100%. It denotes a reduced presence of the C-C stretching vibration of the amino acids proline and tyrosine. Similarly, the DNA/RNA phosphate backbone stretching vibration at 830 cm^{-1} and a feature of the Amide III band at 1284 cm^{-1}, denote the CH deformation of proteins and lipids remain below the 100% mark. Investigating the third and fourth region of interest, a similar trend is notable (Figure 4-8). The CH_2 twisting vibration in DNA / RNA and lipids at 1302 cm^{-1} is consistently lower than the control, although no systematic concentration dependence is observable. The CH stretching vibration at 1342 cm^{-1} displays a similar behaviour. Within these two regions of interest, only the Amide one band at 1656 cm^{-1} is consistently below 100%. The other peak positions vary strongly around and above the intensity of filter control.

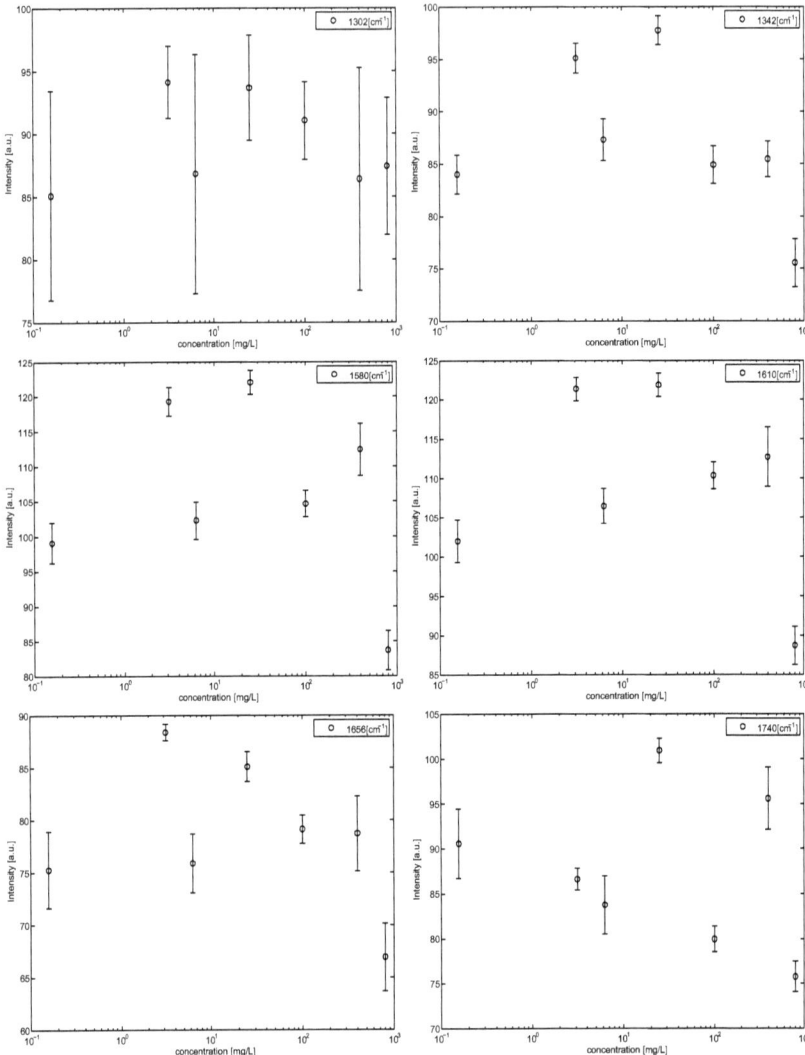

Figure 4-8 Extracts (II) of peak intensity ratios versus CH_2 (1450 cm^{-1}) with the filter control (0 mg/L SWCNT depletion) set to 100 %

Overall, the simplistic analysis of the peak intensity of these regions of interest does not conclusively reveal a response consistent with dose dependent depletion of the cell culture medium. There are some indications of cellular starvation, visibly manifest in some features that are relatable to the overall protein content of the cells, but these do not exhibit systematic concentration dependences. Generally, the markers for cytotoxicity used for the analysis of mercury exposure by Perna et al. seem to correlate only with primary or direct toxicity, also observable in chapter 3. Secondary, or indirect, toxicity of SWCNT, as studied here, essentially is an effect of the depletion of cell culture medium, leading to starvation of the exposed cells, possibly resulting in more subtle effects. For example, medium depletion is a common method to synchronize cells to G0/G1 [184, 185] prior to cell death and does not result in a true toxic effect as such. It should be noted that, 96 hours of cell growth in unreplenished medium leads naturally to nutrient depletion [186], suggesting that even in an unfiltered control, the cells may not be very healthy.

The clonogenic assay, although a gold standard in radiation biology, does not really tell anything specific about the health of a cell or cell colony. It quantifies colony growth in comparison to an unexposed control, which is interpreted as relative viability without any mechanistic support. In the study of Casey et al. [101] little change in the colony number as a function of SWCNT exposure of the medium was observed. The secondary effect was notably manifest in the colony size, however, which was interpreted as a change in the proliferative capacity of the cells. Thus, rather than being directly toxic the depleted environment results in reduced metabolic activity of the cells. Alamar Blue (Resazurin), also employed within the medium depletion study of Casey et al. [101], is more specific, as it indicates proliferative capacity on a metabolic level, although essentially only the redox cascades inside a cell are monitored by this assay [187]. Therefore, no detailed mechanistic response is represented by either assay. In both cases, however, it was seen that significant responses were only apparent at high exposure concentrations [101]. Furthermore, recent studies credit the formation of oxidative stress to resazurin itself, which makes AB further questionable as a reliable marker for the analysis of cytotoxicity [187]. Raman spectroscopic analysis potentially provides a more detailed and noninvasive chemical analysis of local cellular changes and thus multivariate analysis of the spectroscopic data was employed to explore the possibility of modeling the broader cellular response caused by starvation due to medium depletion.

4.3.2.2 Multivariate Analysis

The previously recorded data were fed into the principal component analysis, and analysed for clustering, group separation and any differentiating spectral markers.

The dataset used contained only the samples that were filtered and depleted. The scatter plot, shown in Figure 4-9, reveals a distinct separation between the filter-control and the depleted medium exposed line map averaged cellular spectra. The mean spectra of the depleted medium cells, however, do not cluster according to the exposure dose of the medium before filtering (Figure 4-9).

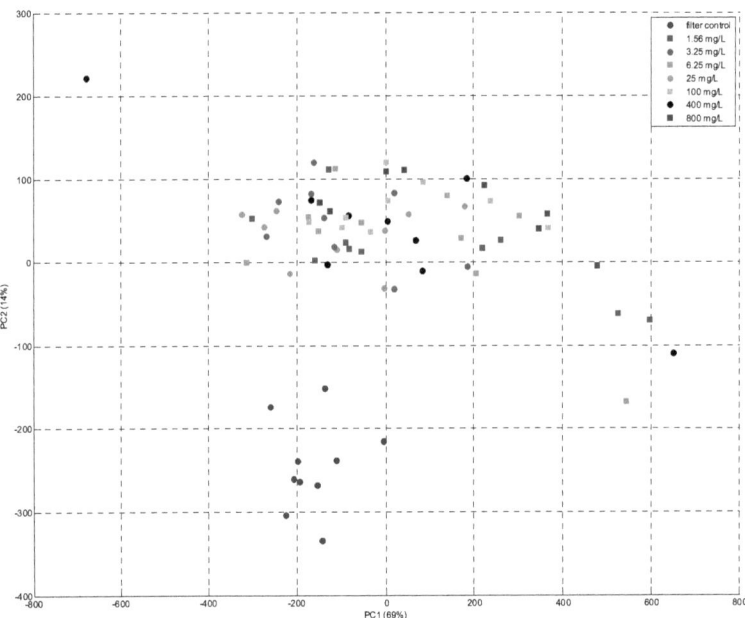

Figure 4-9 Scatterplot of the mean spectra of the depleted medium exposed cells after PCA, with PC1 describing 69% of the total variance and PC2 describing 14% total variance of the dataset.

The elongation or stretching of the cluster of the spectra of the depleted medium exposed cells is predominantly along the dimension of principal component one, which represents 69% of the total variance of the dataset. The separation between the filtered-control and the spectra of the depleted medium exposed cells mainly derives from the second principal component, which describes 14% of the overall variance in the data set.

Prior to the analysis of the loadings of each principal component, it becomes obvious that the medium depletion caused by single walled carbon nanotubes exposure altered the cellular spectra, although it appears that the multivariate nature of the variance is not systematically dependent on the exposure dose. Derivatization of the spectral data prior to PCA did not result in any improved dose dependent classification of the data. Taking the mean density of the data cloud per concentration into account, however, the data suggest that, although hugely influenced by some outliers, with increasing concentration above 6.25 mg per litre, the data sets for each dose become more spread, possibly indicating that the SWCNTs were less well dispersed with increasing concentration [101]. Indeed, it has been demonstrated that, in aqueous surfactant as well as organic solvent dispersions of SWCNTs, the degree of exfoliation of bundles and therefore the variation of aggregation state is strongly concentration dependent [188, 189].

The loadings plots of the first two principal components describe about 84% of total variance (Figure 4-10). They display the first two orthogonal dimensions of the variance of the total dataset and, due to their relation to the spectral sub space, the information can be interpreted in a similar way to spectral data. The first principal component loading represents, roughly, an inverted average cellular spectrum (Figure 3-3(a)). This seems to indicate that the filter-control spectra, which consistently score negatively in PC_1, are spectroscopically closer to healthy cells, whereas most of the spectra of particularly higher doses score positive on PC1, indicating a reduced state of health.

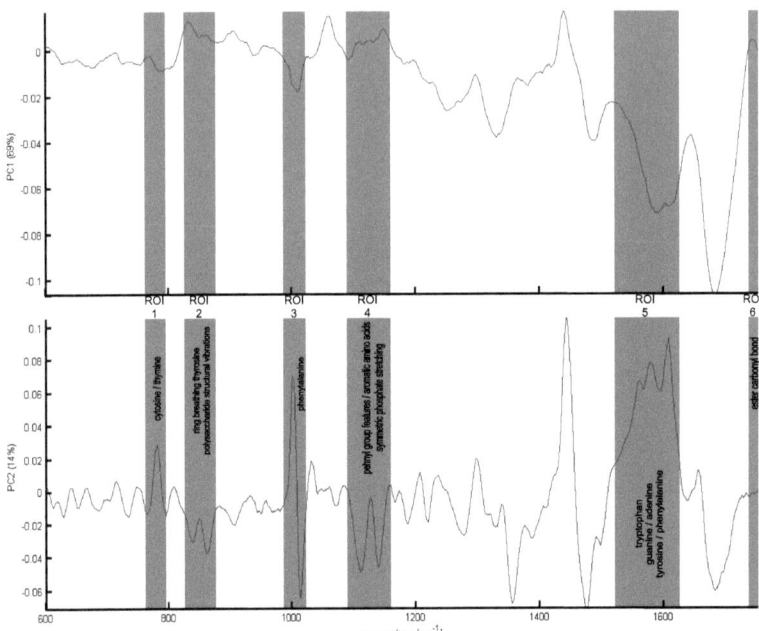

Figure 4-10 Loadings plot of the first two PC's of the depleted medium dataset (PC_1 describing 69% variance and PC_2 describing 14% variance) with shaded regions of interest.

The second principal component loading features some sharp first derivative-like features over the whole spectral range indicating for example a slight shift of the phenylalanine ring breathing at ~1003 cm^{-1}. The variance of the filter-control cells, compared to the depleted medium spectra, is negative in the first and second principal component and thus anti-correlated to PC_1 and PC_2. In terms of PC_1, this indicates stronger cellular spectral features of the filter control cells compared to the exposed cells.

On the other hand, most of the spectra of the depleted medium exposed cells score positive in the first and second principal component. Therefore the variance represented by PC_1 and PC_2 describe the difference between the spectra of the depleted medium exposed cells and the filter-control spectra well, but the separation between both groups is predominantly caused by the different scoring with respect to PC_2.

In PC_2, five regions of interest (ROI) are highlighted due to their inverse occurrence and their intrinsic cellular relevance. These features are already notable, though obscured in the overview of the average spectra in Figure 4-4. Within the first ROI, there is a strong component of variance associated with the feature at ~781 cm^{-1}, which is attributed to DNA-base vibrations of cytosine and thymine [93]. In the second region of interest, 830-863 cm^{-1}, the ring breathing mode of tyrosine and polysaccharide structural vibrations [190] are present in an inverse sense. The phenylalanine feature caused by ring breathing vibrations at ~1003 cm^{-1}, is seen to shift to lower frequencies, which could indicate a biological response to the depletion of the culture medium. Alternatively, this unusually very sharp feature is also very prone to shifts as a result of changes in intensity of overlapping tails of neighbouring broader features of the neighbouring C-O vibrations of RNA Ribose at 960 cm^{-1} and 1056 cm^{-1} and a C-N stretching vibration at 965 cm^{-1} (Table 2-4). This can induce variance in the x direction, which effectively broadens or shifts the phenylalanine feature and phenyl group features of any of the aromatic amino acids. In the spectral region between 1100-1150 cm^{-1}, the negative occurrence of the symmetric phosphate stretching vibration at ~1110 cm^{-1} and the inverse appearance of the saccharide stretching vibration at -1139 cm^{-1} are obvious. The last region of interest in the second PC-loading is the spectral region just before the Amide I band. Between 1550 cm^{-1} – 1610 cm^{-1} lies the inversely varying peak attributed to tryptophan [190] and, similarly oriented, the nucleic acid attributed features of guanine and adenine at 1578 cm^{-1} and the tyrosine and phenylalanine contributed ring breathing vibration at 1607 cm^{-1}. The latter two have already been identified as markers for cell death in previous studies [191]. Another marker for cell death along with the Amide I and Amide II bands is an increase in the ester carbonyl bond vibration at ~1743 cm^{-1} [192] which is present in the loading of the first PC. Projecting this information onto the negative scoring of the filter controls in the second dimension generally suggests a better preservation of the cells after fixation than the exposed cells. By extrapolation, as the filter controls score negative, the first areas of interest might be understood in terms of better 'health' before fixation. Although it has been reported that the fixation of A549 cells with formalin has minimal impact on the spectra [182], the fixation step during sample preparation has some impact on the spectra and therefore probably not all previous markers can be employed as markers for cell death due to starvation as such. Though all of the spectra show this behaviour, it is obvious that the spectra of the exposed cells are more strongly affected.

Finally, the unfiltered control and filter control spectral data were subjected to PCA. Surprisingly, the filtering of the unexposed medium itself had an effect on the clustering of these two control sets (Figure 4-11) represented here by their (line-map) means. Predominantly the datasets are separated along the first principal component. The corresponding loading (Figure 4-12), although inverted, displays a number of similarities to the previous loading of principal component two (Figure 4-10), whereas markers for dead cells at for example 1587, 1607, 1743, cm^{-1} [93, 191, 192] are not in evidence.

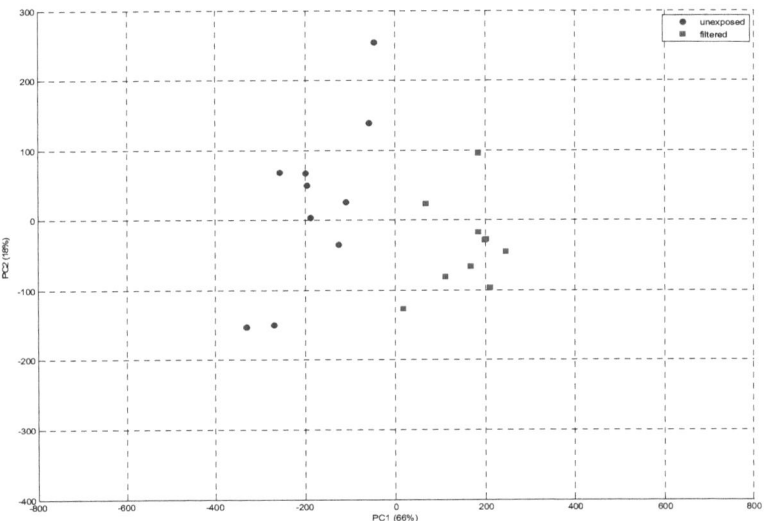

Figure 4-11 Scatterplot of the mean spectra of the control and filter control medium exposed cells after PCA, with PC$_1$ describing 66% of the total variance and PC$_2$ describing 18% total variance of the dataset.

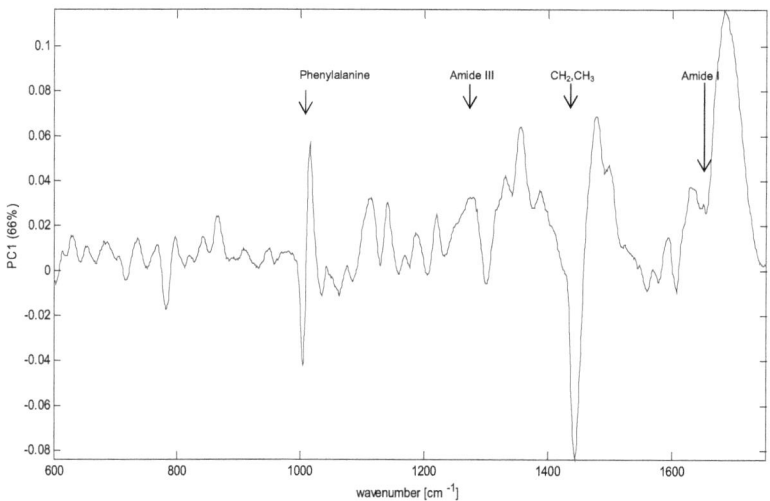

Figure 4-12 First PC describing 66% of variance of the control and filter control medium exposed cells after PCA.

Thus, the filter control cells score positively throughout. It can be assumed that the filtering process itself has an influence on the medium constituents. This links the filtering process in both PCA to the outcome separation between the groups, indicating that the filtering process has an influence on the cell culture medium. As the first principal component does not show any similarities with a common spectrum of A549, the influence on the health of the analysed cells is not fatal. Whereas the specifically depleted medium created variance in the PCA of Figure 4-10, the most variance was captured along a negative cellular spectrum. Therefore, filtering of the medium clearly changes the spectral characteristics of the cells after 96h but the changes are less prominent than those in the SWCNT induced depletion. The health of the cells exposed to the filtered medium is different but comparable to the health of the unfiltered controls, although significantly different to the SWCNT exposed.

Filtering the medium appears to deplete the medium to a small extent, although in terms of the protein content, this depletion appears to be minimal (Figure 4-3). Exposure of the medium to SWCNTs and subsequent filtration results in a significant, dose dependent medium depletion, to a maximum concentration of ~100mg/L. However, over a 96hr period, the unexposed cells deplete the medium as a result of their natural metabolism, and it is likely that this natural depletion of the medium occurs at rates which are dependent on the initial state of the medium. Such a process can be illustrated schematically as in Figure 4-13 which suggests how the change in health of the cells due to exposure to depleted medium might evolve with time. Despite the fact that all cultures suffer from medium depletion to different degrees, after 96 h, the difference may be minimal.

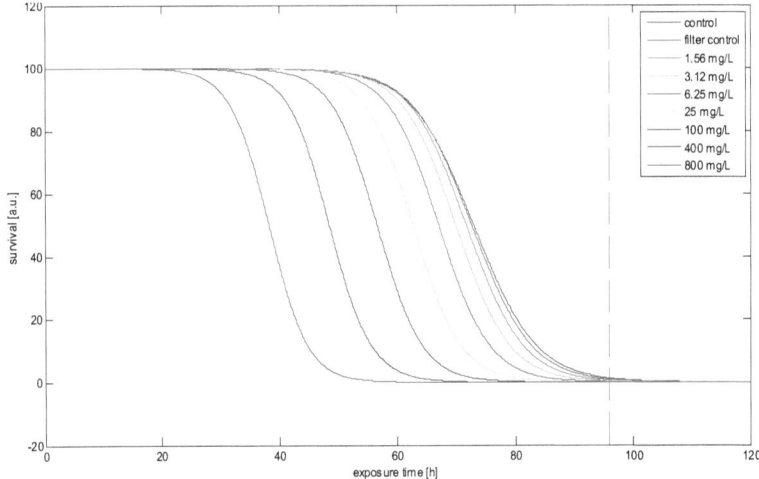

Figure 4-13 Illustration of suggested viability response in dependence of starvation time and intensity expressed in % of viability (96h time point highlighted).

Reduction of the proliferative capacity of a cell culture by starvation as a result of medium depletion is known to result in cell cycle arrest [184, 185] or apoptosis [186]. Noting the markers for changing protein content and poly saccharide alteration [103], it can be understood that all cells including control suffered some degree from nutrient deficiency over 96h of exposure, but the cells exposed to the SWCNT depleted medium were probably starved earlier, and thus the signals of death are more prominent. This argument supports the depletion aspect of the experiment, although no direct concentration dependence is apparent. Similar to the toxicological study, only the very high concentrations above 100 mg/l were seen to indicate a significant result in the colorimetric analysis with Alamar Blue [101].

In an additional attempt to uncover a possible but unlikely dose dependent response, the dataset was fed into a partial least squares regression, employing the Alamar Blue assay data for 96 h, obtained by Casey et al. [101], as the regression targets, serving as markers for starvation induced by culture medium depletion due to

SWCNT exposure (Figure 4-14).

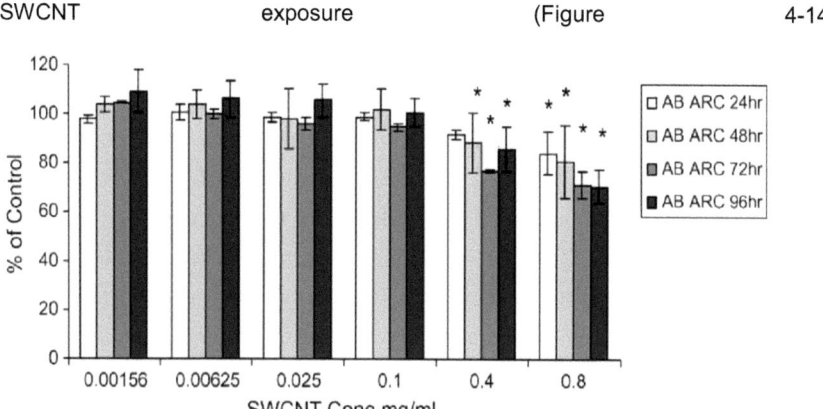

Figure 4-14 Cytotoxicity of Arc Discharge SWCNT filtered medium to A549 cells after 24, 48, 72 and 96 h exposures determined by the AB assay. Data are expressed as percent of control mean ± S.D. of three in dependent experiments. (*) Denotes significant difference from control (p≤0.05). (taken from Casey et al., 2008)

The possibility of building a simple PLS regression model with the full dataset as training and cross validation set, finally presenting the unknown concentration of 3.12 mg per litre to the model, was investigated. Therefore, the dataset was reduced to 748 spectra by removing 99 spectra (3.12 mg per litre), each with 1153 data points. For the initial construction of the model, in order to obtain a qualified number of latent variables, 100 % of the mean centred dataset were used. The calibration model was rigorously cross-validated. The optimal number of latent variables was found to be 10 out of 20 possible, with a maximum R^2 of 0.39, and consequently the modelling of the data via PLS regression was abandoned due to the maximum improvement of R^2 to 0.59 by employing 20 LV's, falling short of a good result with an R^2 above 0.95. Employing high numbers of latent variables additionally makes the model prone to severe over fitting which further removes actual predictive capabilities. This lack of additional explanatory power supports the initial impression, indicated by PCA, in which no clear dose dependent indirect response is observable. Indeed, it is notable that the Alamar Blue results of Casey et al. only indicate a significant metabolic reduction due to starvation at the highest exposure doses.

4.4 Discussion and Conclusions

Although Raman spectroscopy, in combination with multivariate analysis, has proven to be a valid tool to assess and analyse cytotoxic responses [170], in the present study it is shown to have had limited utility. In the univariate analysis, it becomes apparent that the statistical variations in the spectral data are very large. Whereas in the study of chapter 3 much of the variation can be attributed to insufficient dispersion of nanotubes, Figure 4-3 indicates that the medium depletion because of SWCNT exposure was systematically dependent on exposure dose. The filtration process itself was seen to minimally impact on the protein content of the medium, as represented by the strength of the Amide I band compared to the control. Nevertheless, this small change in the medium content appears to influence the state of health of the cells after 96 h exposure, as indicated by the PCA of Figure 4-11. The spectral loading of the PC, which differentiates the filter control and the control, also differentiates the filter control from the cell cultures grown in the SWNT exposed media, indicating that this multivariate fingerprint is a signature of cell starvation. However, no systematic dose dependence is observed. The increasing spread of the data sets according to PC_1 in Figure 4-9, indicates that although the overall protein content may be systematically dependent on exposure dose, as in Figure 4-3, more subtle variations in medium composition may result from variable dispersion of the SWCNTs in the exposure medium dispersions.

The data spread according to PC_1 in Figure 4-9, indicates variations in the state of the cellular health, as represented by the loadings plot of Figure 4-10. In the clonogenic study of Casey et al., medium depletion appears to affect the proliferative capacity of the cells rather than the viability of the culture. The Alamar Blue assay confirms the effects on the cellular metabolism. It appears that such effects are not manifest spectroscopically as distinct cytotoxicity markers, as observed by Perna et al. for the case of mercury toxicity [107] and in chapter 3 for the case of the toxic response as a result of direct exposure to SWCNTs.

In comparison to controls, the effect of incubation for 96 h in filtered medium is manifest in the spectral profile of PC_2. This spectral signature also differentiates the cultures grown in SWNT exposed media from the filtered controls, although not in a dose dependent fashion. It appears therefore that this signature is an extremely sensitive marker of cell metabolism in the early stages of exposure to cell depletion.

While the clonogenic and AB assays only differentiated the high dose exposures, Raman spectroscopy, with the aid of multivariate analysis, can differentiate the effects of weakly depleted media. It appears, however, that the effect determined is saturated at higher dose points. This may be because of the duration of the exposure, at the end of which the medium is significantly depleted even under normal cell growth conditions.

PLS regression failed to establish a model of the spectroscopic signatures as a function of exposure dose or metabolic response, as determined by the AB assay. The PLS regression process is a linear model and so assumes a linear relationship between the multivariate features and the target. Given the indications that the spectroscopic response saturates at relatively low doses, the inability of PLS regression to construct a statistically viable model may not be surprising. Notably, it has previously been observed for the case of the effect of ionising radiation *in vitro*, that the physiological response as measured by the AB assay is quadratically related to the variation in identified spectroscopic peak ratios [Aidan D. Meade, PhD Thesis, DIT, (2010)]. Inversely, the spectroscopic response is related to the square root of the assay response and so while sensitive at low doses, saturates at high doses.

It is also noted that the univariate and multivariate analyses were performed on spatial averages over the cytoplasm and nucleus. Changes in metabolic activity resulting in changes in proliferative capacity are most likely initially manifested as changes in the cytoplasm. However, a spatial average will be dominated by the features of the dense nuclear region and so may not best represent the subtle changes associated with secondary toxicity. A more sensitive study could probe the effects over a shorter exposure, timeframe concentrating only on the cytoplasmic components.

Chapter 5 : Study of toxic responses as a result of Reactive Oxygen Species

5.1 Introduction

Nanoparticle toxicity has been proposed to have origin in the induction of oxidative stress by free radical formation [1, 43, 66, 193, 194]. SWCNTs result in increased levels of Reactive Oxygen Species (ROS) in cell cultures, as previously demonstrated for A549 lung cells [1]. ROS are secreted as a first immune response of cells to threatening stimuli such as for example bacteria [46] and are a general cellular stress messengers [195]. They are also generated and neutralized in peroxisomes during metabolisation of fatty acids and by highly specific enzymes [47]. Excess amounts of ROS are however known to have a strong potential to damage cells [196]. ROS easily migrate through cellular membranes [46] and are known to have various detrimental effects on cellular viability [197]. They can induce DNA damage by modifying the desoxyribose-complex or individual nucleic bases, leading to strand breaks or causing deletions and oxidative modifications. In proteins, drastic changes can be induced by ROS, including limitation of the activity of elastase, which is connected to the development of lung emphysema [196]. Lipids, especially unsaturated lipids, are prone to lipid peroxidation [198], which in itself can become autocatalytic [46]. Oxidative stress may have a role in the induction of inflammation through up regulation of redox sensitive transcription factors, NF-κB and activator protein-1 (AP-1), and kinases involved in inflammation [199, 200].

In order to further explore the potential of Raman spectroscopy as a toxicological screening tool, the increased ROS levels produced by exposure to nanoparticles are mimicked by a time dependent exposure to hydrogen peroxide at the cytotoxic level of $\geq$ 50 µM [201, 202]. Hydrogen peroxide is a strong oxidant that generates ROS in Fenton like reactions with for example metal ions [197, 203] and is a byproduct of cellular metabolism, as well as being a cytokine in its own right [44, 46]. It will be demonstrated that Raman spectroscopy is capable of picking up the different responses of A549 lung cells as a result of exposure to hydrogen peroxide *in vitro*, and a correlation to the mode of action by univariate and multivariate analysis will be established. The concentration of hydrogen peroxide applied in this study is that of the reported critical level of cytotoxicity $\geq$ 50 µM [201, 202]. Therefore, it is assumed that this level is an appropriate sensitivity level for Raman spectroscopy as a probe to access cytotoxicity.

5.2 Materials and Methods

5.2.1 Sample preparation

A549 cells were cultivated as previously described in section 2.4.3, and harvested at ~85% confluence. As substrate for spectroscopy, round, surface modified (polished) quartz slides (Crystran Ltd. Poole, UK) were washed with hydrochloric acid and ethanol and air dried under laminar flow, then loaded into six well plates (Nunc A/S,

USA) and covered with 3 mL cell culture medium as described in chapter 2.4.5. The chambers then received 5×10^4 cells, each, in the center of the quartz substrate and were incubated for 48 hours to allow the cells to attach to the substrates. After 48 hours, the medium was removed and the cells were washed with PBS three times before exposure. The exposure solution consisted of cell culture medium with hydrogen peroxide added at a final concentration of 50 μM. The control cells received unmodified cell culture medium. After one, three and six hours of exposure, the slides were rinsed with PBS and then fixed in 4% formalin in PBS solution for 10 minutes, and subsequently rinsed three times with deionised water. They were finally stored, immersed in dH_2O at 4°C, before the measurements were conducted. The control cells were fixed in the described manner six hours after medium change.

5.2.2 Spectroscopy

Raman Spectroscopy was carried out with a Labram 800HR Raman confocal microscope using a 532 nm (frequency doubled Nd^{3+}: YAG) laser as source with a grating of 300 l/mm, providing a spectral dispersion of about 1.43 cm^{-1} per pixel. Cellular spectra were recorded using a water immersion lens (Olympus Lum-Plan FL x100) from elevated substrates immersed in water in a sealable Petri dish with an opening for the lens to prevent evaporation of deionised water. The x100 water immersion objective produced a spot size of approximately 1μm in diameter at the sample. The system was calibrated to the spectral line of crystalline silicon at a constant room temperature of 21°C. The recording window was set to ~348-1752 cm^{-1} in order to detect spectra within the fingerprint region of the cell samples [91]. Each measurement was integrated over 90 seconds at a measured laser power of 37 mW at the sample. The substrate spectrum per slide was measured before and after acquiring the cellular spectra, in order to have an indicator for any possible spectral shift during the measurement. Eleven acquisition points per cell were defined, evenly spread along a straight line across the nucleus and the adjacent cytoplasm in order to capture co-localised changes of spectra. All recordings were performed as an average of three individual measurements of one point to reduce the influence of spectral noise.

5.2.3 Data analysis

In total, 528 spectra were recorded, arranged in groups of [4 4 4], separately, in order to afford sufficient samples prior to outlier detection. The preprocessing regime applied comprised of the same steps as used in the depletion study of chapter 4.

First, data were loaded into Matlab 7.3. After linearisation of the substrate recordings, the background was removed with the rubberband function. Any drift between spectral measurements was analysed by cross correlating the pre- and post-recorded scaled substrate spectra per slide. Where drift was present, it was linearly interpolated over the time between the measurements and then integrated into the sample spectra by shifting the abscissae accordingly. Afterwards, the background of the sample spectra was removed with the rubberband function and finally the spectra were linearised. The substrate spectra, scaled to the characteristic quartz feature at 486 cm^{-1} [144], were then deconvoluted by a number of combined Gaussian and Lorentzian functions to model the substrate as described in section 2.4.8.7.3. This model of the substrate was then calculated per scaled spectrum and subtracted. As all measurements were recorded in water immersion and the contribution of water inside the confocal volume might be variable, the Amide I region at about 1656 cm^{-1} was deconvoluted using a series of combined Gaussian and Lorentzian functions, to remove the water contribution at about 1640 cm^{-1}. It was decided not to crop the spectral window *ab initio*, possibly leaving substrate residues caused by the variance of the quartz feature and the influence of the varying Rayleigh scattering in the recorded spectra. Thus, the performance of the selected preprocessing showed no influence on the cellular spectra. Outlier detection was again performed manually after the preprocessing of the spectra by analysing each mean spectrum per cell and 440 spectra were found valid for analysis (Table 5-1).

Table 5-1 Sample numbers (ROS study) after recording the measurements and outlier removal

Sample ROS (50µmol) Exposure	Recorded Replicates			Validated Replicates		
	I	II	III	I	II	III
Control	44	44	44	44	33	33
1h	44	44	44	33	44	33
3h	44	44	44	33	33	44
6h	44	44	44	33	33	33

5.3 Results

5.3.1 Visual Changes

In comparison to the control samples, the microscopic images of the A549 cells which were exposed to the H_2O_2, showed significant morphological changes (Figure 5-1). Whereas in the control cells the nucleus plus the nucleoli are clearly visible, in the exposed cells a clear distinction is not possible. The exposed cells seem to be packed with granules or small spherical vesicles. The cells appear more contracted and the cross section appears to be smaller. Overall, the exposed cells can be immediately identified as they have an abnormal morphology after exposure to hydrogen peroxide. Even though the samples were washed after fixation, the clear presence of cellular debris, visible outside of the cells, gives an indication of induced membrane instability and the tendency to rupture even in a fixed state. Visual inspection does not identify any distinct exposure time dependent responses, but clear signs of intracellular oxidative stress induced by externally applied hydrogen peroxide are evident.

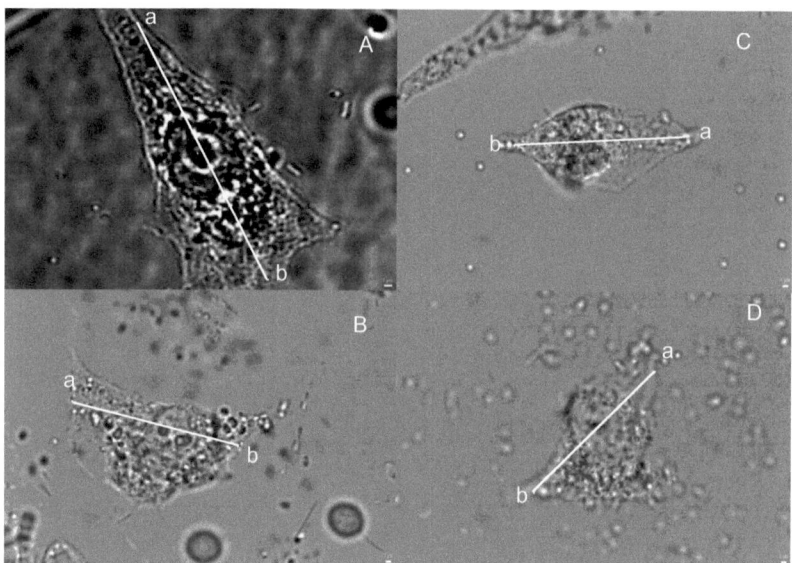

Figure 5-1 Microscopic images of A549 cells exposed to 50μM H_2O_2 for different intervals (A=0h,B=1h,C=3h,D=6h). The overlaid lines indicate the positions of the spectral line maps measured from a=1 to b=11

The cellular responses evident in Figure 5-1 are well described in literature, as the intracellular H_2O_2 is detoxified in peroxisomes [47] inside a cell suffering from oxidative stress [197]. The primary way in which the cell detoxifies external agents is to envelop them in intracellular vesicles. Thus the high number of vesicles observed in Figure 5-1 may be explained [46].

A similar observation was made by Davoren et al., whereby exposure to SWCNT dispersions led to an increased number of intracellular surfactant storing lamellar bodies [43, 75]. Spectroscopic measurements were recorded, as line maps, along the radial axis of each cell, as indicated in Figure 5-2, and therefore provide additional collocational information. Figure 5-2 shows a control cell microscopic image, indicating the regions of the spectroscopic line scan and the corresponding eleven spectra along the line scan. The line map shows the characteristic features of A549 cells in the denser regions of the nucleus, roughly in the center of the map, dominated by the Amide I band at $1590 cm^{-1}$. Notable are the weak spectral contributions in locations of membrane away from the centre of the cell. A spectral average of the line scan will thus be dominated by the spectra recorded in the region of the nucleus.

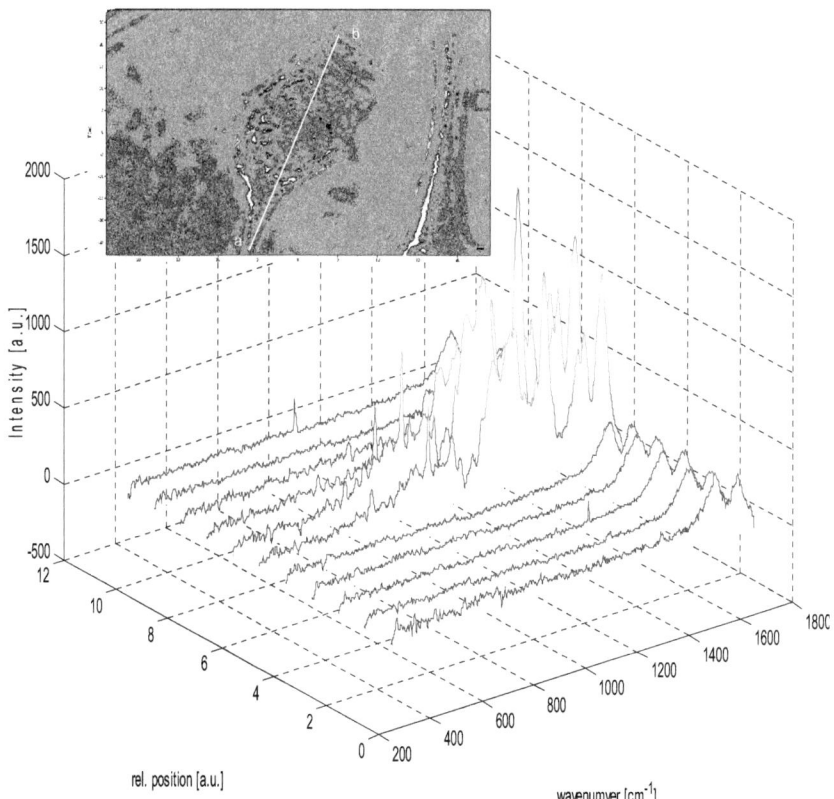

Figure 5-2 Line map from (a) to (b) of an untreated, fixed A549 lung cell. The corresponding microscopic image indicates the direction and positioning of the line scan on the cell

After only 1hr exposure of the cells to H_2O_2, however, the spectral profile of the cell changes significantly. There is now no discernible transition from membrane to nucleus. Strong spectral features in the region of the Amide I band are seen across the cell, and their intensities are relatively higher, than those of the nuclear region of the unexposed cells. As the scale of the images are similar, it is suggested that the cellular cross section is reduced upon exposure to ROS, which might be explained by the increased rounding of the cell membrane, a known strong effect in A549 cells exposed to oxidative stress [204]. This rounding up of cells changes the strength of adhesion to the extracellular matrix, here the quartz slide, and alters the spatial distribution of cell organelles inside the cytoplasm. In the process of ROS induced cell death strong vesicularisation or cell blebbing is observable. Prior to membrane rupture, cell organelles and fragments of the condensed nucleus are encapsulated in membranes, which might further decrease the cell membrane surface area and increase the rounding up of the exposed cell, leading to possible rupture [205] and subsequent cell death.

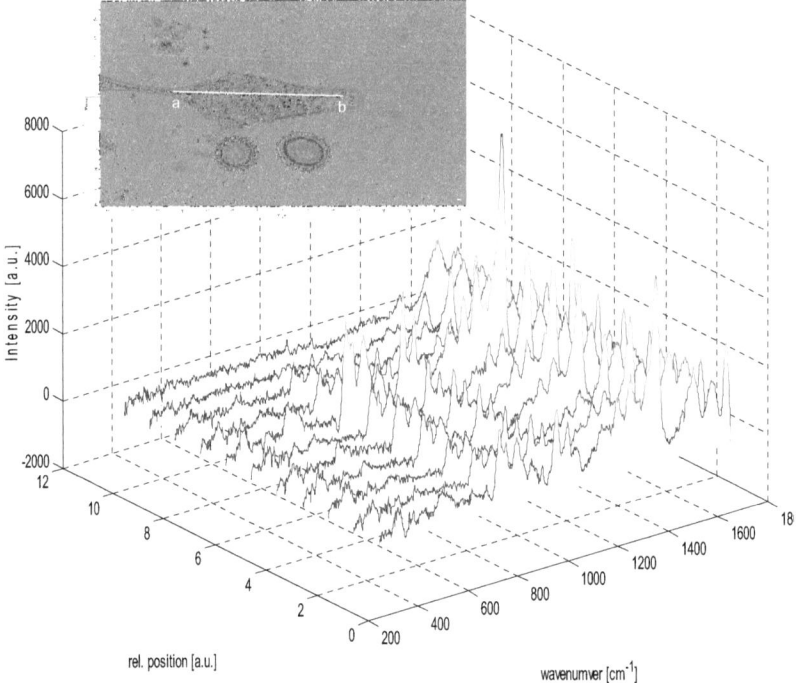

Figure 5-3 A549 cell exposed to ROS for 1h (line map) and the corresponding microscope image, illustrating the location of the spectral map.

5.3.2 Spectroscopic Changes

The line averaged spectra of the three different exposure durations plus control show significant differences (Figure 5-4). These differences are manifest as a series of sharp peaks, so pronounced that they could be noted already during the measurement and could have been mistaken as artefacts. These appear in the lipid, protein and DNA attributed regions of the average spectra of ROS exposed A549 cells.

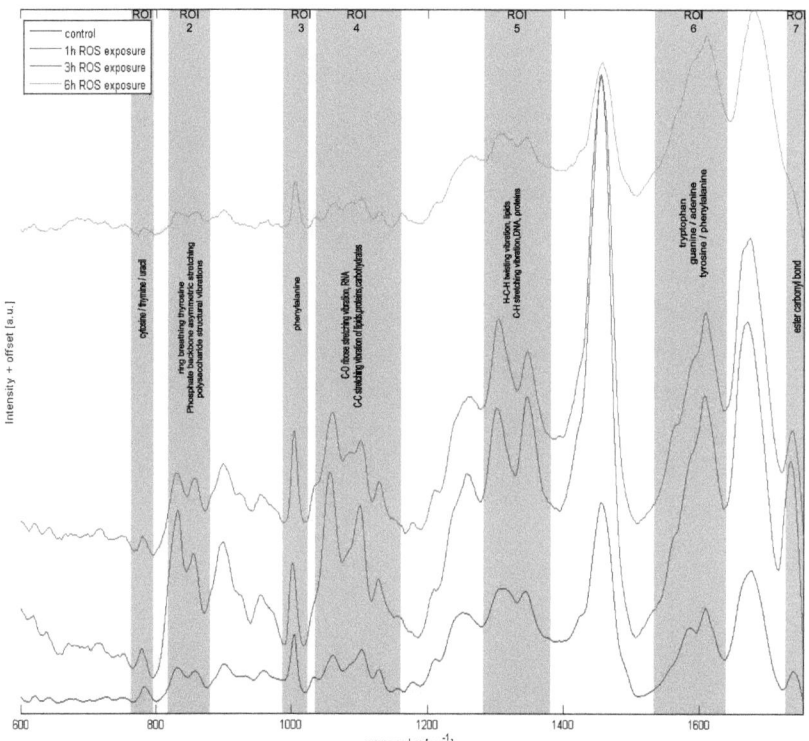

Figure 5-4 Average cellular spectra of ROS (induced by H_2O_2) exposed A549 lung cells (control, 1h, 3h, 6h)

In detail, seven different regions of interest can be easily identified. Left to right, the first feature is the ring breathing vibration of RNA at ~782 cm^{-1} (thymine, cytosine, uracil) [129]. The second feature cluster includes the out of plane ring breathing vibration of proteins and the phosphate backbone asymmetric stretching vibration of DNA & RNA at 832cm^{-1}. In the third ROI, the ring-breathing mode of phenylalanine at 1003 cm^{-1} features strongly. The fourth cluster comprises of the C-O stretching vibration inside the ribose of RNA at ~1056cm^{-1} [91], the phosphate symmetric stretching vibration at 1100 cm^{-1} of nucleic acids and the C-C stretching vibration of proteins, lipids and carbohydrates at ~1126 cm^{-1}. Another lipid attributed feature set is prominent at ~1302 cm^{-1} (CH_2 twisting vibration) and the feature at ~1342cm^{-1} originates from poly nucleic chain vibrations in DNA and C-H stretching vibrations of proteins [129]. The last two ROI's feature the strong C=C bending vibration of proteins at ~1607cm^{-1} and the C=O (ester) stretching vibration in lipids at ~1740 cm^{-1} [190].

It seems that, upon exposure to hydrogen peroxide, between 1-3 h, the overall intensity of the spectra increases, whereas after 6 h this intensity seems to drop to control levels. The spectrally integrated intensity of the measured spectrum increases nearly 2.5 fold for 1 h exposure whereas for 3 h exposure the intensity increase is twofold, finally dropping to nearly the original values of the control after 6h exposure, as shown in Figure 5-5. Notably, the intensity variation of individual features such as the ring breathing vibration of the Phenylalanine ring at 1003 cm-1 show similar, although not identical, responses (Figure 5-6). It seems that the A549 cells respond strongly during the first three hours of exposure and return to a state, which is spectroscopically similar to the control after six hours. This is manifest as strong features that seem to initially increase and subsequently decay with increased exposure duration.

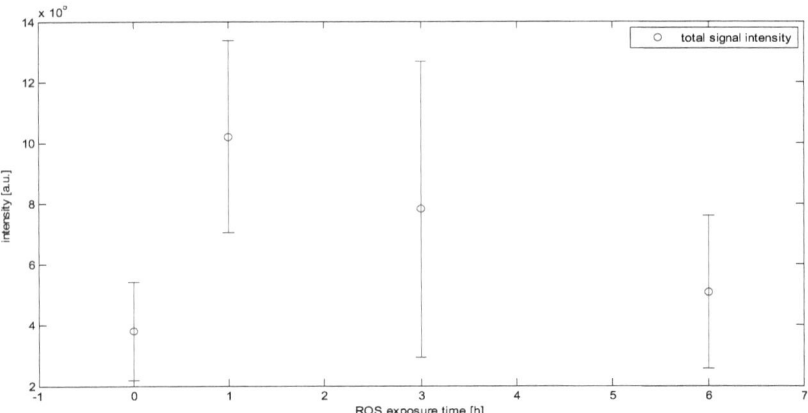

Figure 5-5 Integrated signal intensity response of A549 in time dependence to the exposure to ROS

As described in the previous chapter, the feature of CH2 at ~1449 cm-1 may be attributed to bending and scissoring vibrations in proteins and phospholipids [105, 182], and as CH and CH2 are the most common groups inside a biological compound [183] the feature can be used to normalize the spectra for their overall biological content [93]. Notably, this feature displays similar effects (Figure 5-7) to the integrated intensity of ROS exposed samples (Figure 5-5) and therefore it is difficult to justify its use as a scaling parameter.

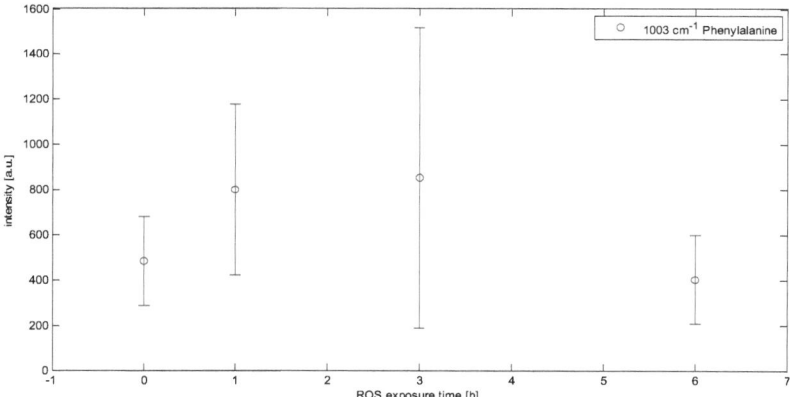

Figure 5-6 Intensity of the fitted Phenylalanine response of A549 in time dependence to the exposure to ROS

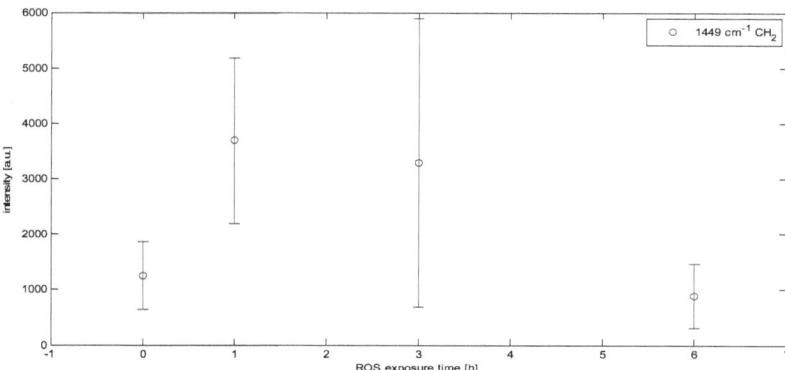

Figure 5-7 CH_2 feature at ~1449 cm^{-1} (fitted) in dependence of ROS exposure time.

Considering the co-locational aspects of the CH2 feature at ~1449 cm-1 along a spectral line map of short-term exposed cells (1 h), its intensity (among others) is strongly varying in comparison to that of the control (Figure 5-2, Figure 5-7). The features that were eye catching in the spatial averages are evident in the spatial profile, but are not limited to a specific location. Moreover, as illustrated in Figure 5-3, these features are present along the extent of the line map, indicating a response that takes place in or close to the cellular membrane.

A potential explanation of the observations is lipid-peroxidation, after Fenton–like reactions, in the cell membrane, due to the ease of abstraction of single hydrogen atoms in an unsaturated fatty acid e.g. oleoyl chain of one membrane phospholipid [46]. Auto peroxidation is a possible result that leads to the creation of vast numbers of lipid-peroxides wherever a ROS is capable of abstracting a hydrogen atom from an unsaturated fatty acid. Secondly, as catalase and superoxide dismutase (SOD), cellularly located in peroxisomes [46], are the active form of the ROS reduction response of cells, it is quite possible that the presence of primarily extracellular ROS triggers the peroxisome biogenesis genes [195], thereby inducing DNA associated spectral changes, possibly represented by the spectral changes in the DNA and RNA attributed ROI's. This may be due to actual damage of DNA and its ongoing repair process. Before lipid peroxidase can be degraded in peroxisomes, a process which is not coupled to ATP generation, the fatty acids are esterified to fatty acyl coenzyme A [206], possibly explaining the rise in the last spectral ROI at ~1740 cm^{-1} (Figure 5-8), representing the ester bond (C=O) [207]. The large error bars indicate the strong variance in response to ROS threat and the collocational aspects of this response.

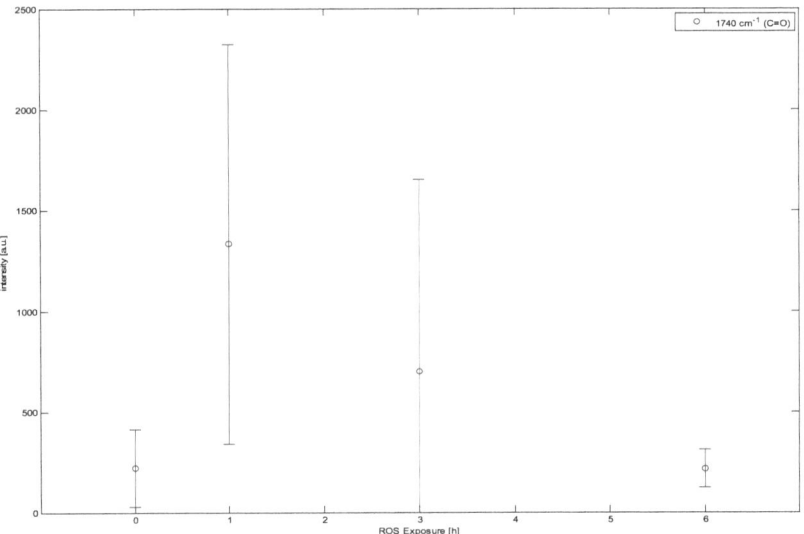

Figure 5-8 Variation of Ester bond at ~1740 cm-1 as a function of ROS exposure time

Previously identified markers of cytotoxicity within the Amide III [107] band, namely the CH2 twisting mode vibration at 1302 cm-1, the Amide III α-helix attributed CH bending vibration, at ~1287 cm-1 and the CH deformation vibration at ~1338 cm-1, normalized to the intensity of the Amide III random coil feature at ~1238 cm-1 do not show a clear decrease in intensity with exposure time, as shown in Figure 5-9.

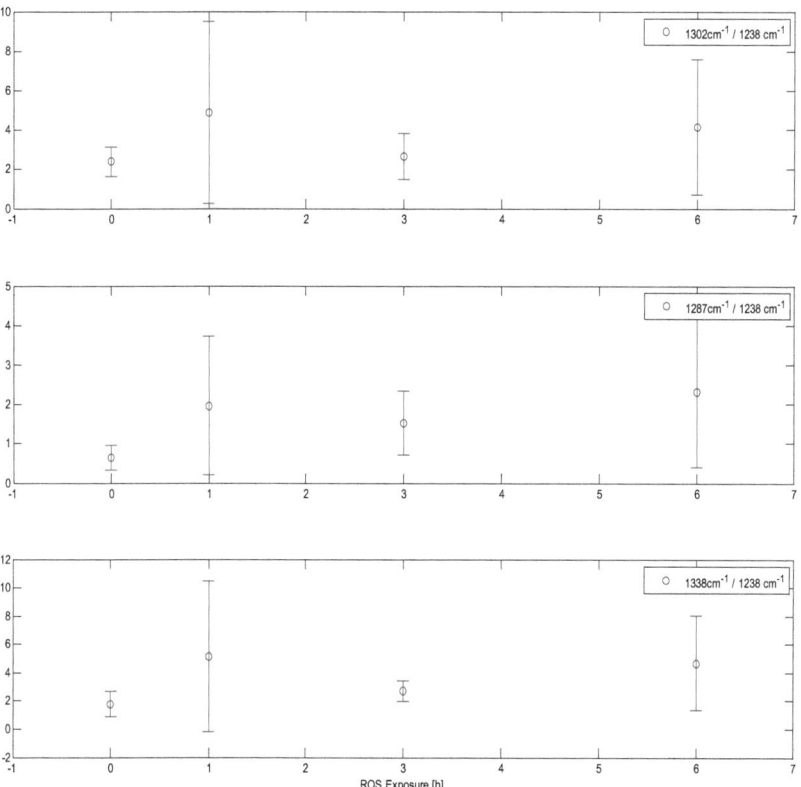

Figure 5-9 Variation in time of toxicity markers (from Perna et. al.) of A549 lung cells exposed to 50 µM ROS (0-6 h)

The ratios display similar results for the control and 3 h exposure and similar results for the 1 h and 6 h exposure. The lack of correlation with the identified toxicity markers, as in the case of the medium depletion in the previous chapter, can be understood in the sense of ROS not being toxic, in a biological sense, but having a strong influence on certain chemical moieties (e.g. CH_2) in lipids and proteins. Alternatively, the lack of correlation could be due to the masking influence of strongly varying neighbouring features, which leak into the Amide III spectral region. Overall, the complex spectral variations observed are difficult to resolve by univariate analysis, but provide a promising starting point for subsequent multivariate analysis.

5.3.3 Multivariate Analysis

In a multivariate approach, all of the ROS exposed spectral data were subjected to principal component analysis after preprocessing. The preprocessing, similar to that used for the univariate analysis, consisted only of substrate and water removal by fitting with multiple Gaussian and Lorentzian functions and manual sample outlier rejection. With the assumption that the major effects of reactive oxygen species take place all across a line map of the cell, no spatial averaging was applied, and thus each measured spectrum, delivers one data point.

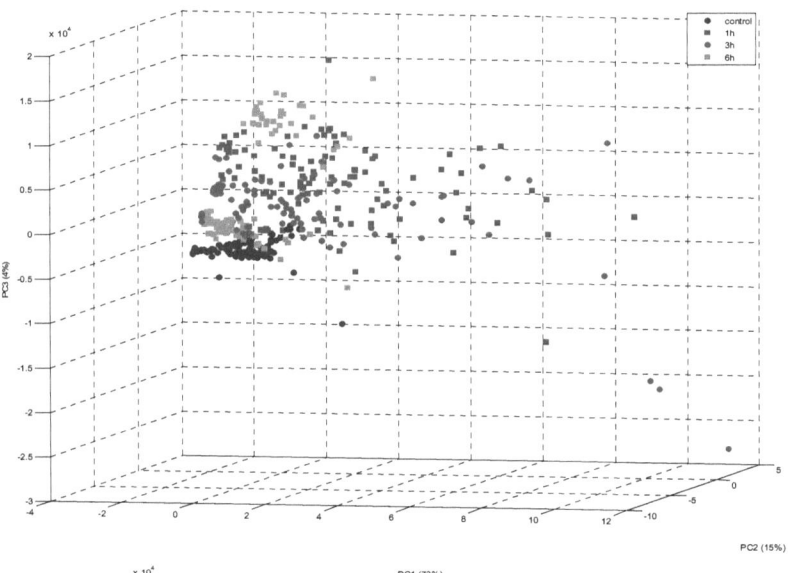

Figure 5-10 Scatterplot of scorings of the first 3 PC's

The first three principal components represent a total of 92% of the variance, the first 73%, the second 15% and the third 4% of the total variance. The scatter plot, colour-coded for each time point, (Figure 5-10), does not show a clear separation of distinct clusters associated with exposure times. It shows a dense cloud of data points for the control group and four differently scattered data clouds. The data points representing the six hour exposed samples are grouped in two distinct clusters. One of them is located close to the cluster of the control samples and is relatively dense. The data points of the 1 h and 3 h exposed samples, as well as the second group of 6h exposed samples, are diffusely spread predominantly along PC1, and are separated from the control and the first group of 6 h exposed samples by PC_3. Principal component two, does not contribute to this slight separation, but broadens the data cloud of the 1 h and 3 h exposed samples, as shown in Figure 5-11.

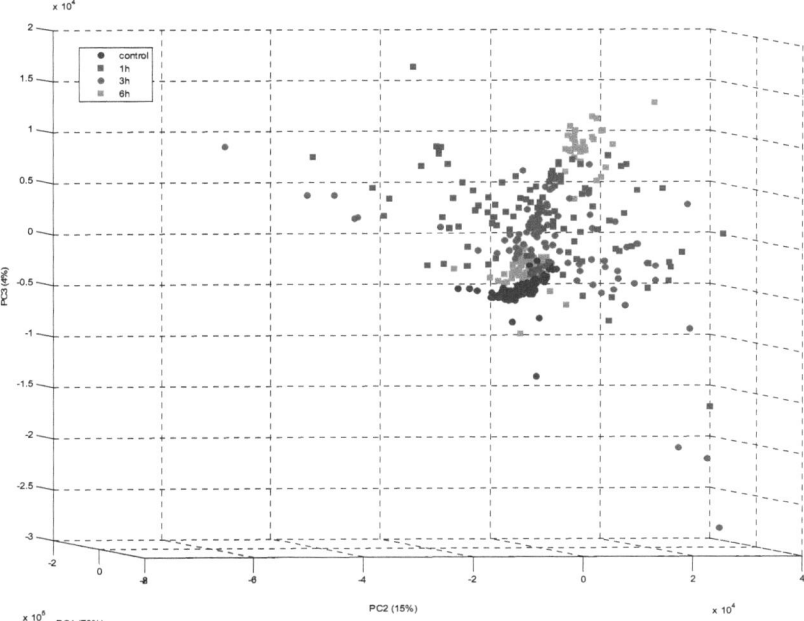

Figure 5-11 Scatterplot of scorings of the first 3 PC's rotated around the y-axis

The data points representing the samples that were exposed to reactive oxygen species for one and three hours describe a much larger and less dense data distribution which is not only separated along the third principal component but predominantly along the first principal component. Notably, the majority of samples that were exposed for one-hour score higher on the third principal component than the samples that were exposed for three hours, whereas the samples that were exposed for 6 hours score similar to the controls, though slightly higher.

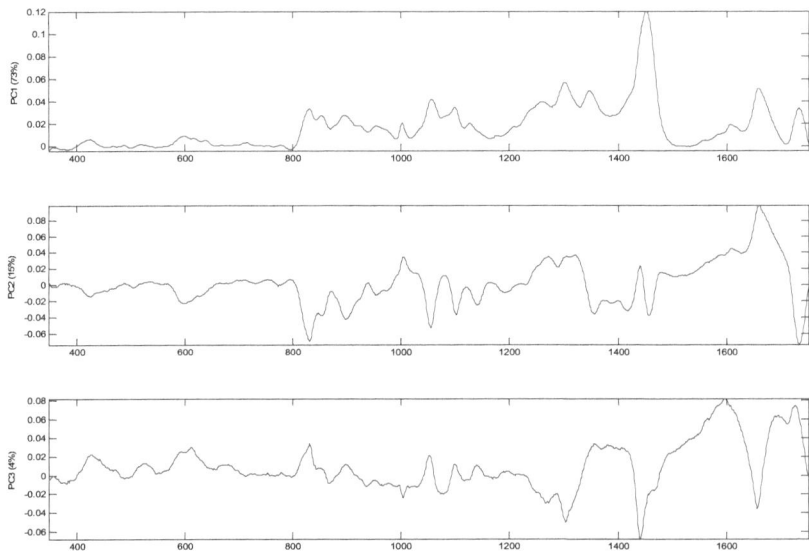

Figure 5-12 Principal component loading plot of PC_1, PC_2 and PC_3 of ROS exposed A549

None of the loadings of the first three principal components displays features similar to a typical cellular spectrum, although they display variations that were previously identified in the univariate analysis (section 5.3.2). This implies that the variances of the typical cellular spectral features are minor and are obscured by the strong changes induced by ROS. The close co-localisation of many of the 6 h exposed spectra indicate that these recover almost completely to the level of the controls.

The spectra of the 1 h, 3 h and the second grouping of 6 h exposed samples are primarily separated by PC_3. The loading of PC_3 shows considerable resemblance to that of common intracellular lipids, such as phosphatidylcholine, as shown in Figure 5-14. The dominance of such features in the principal components of the variance is consistent with the observed increase in lipidic bodies in the micrographs of Figure 5-1. It is furthermore understandable that the features are distributed across the spatial extent of the cell.

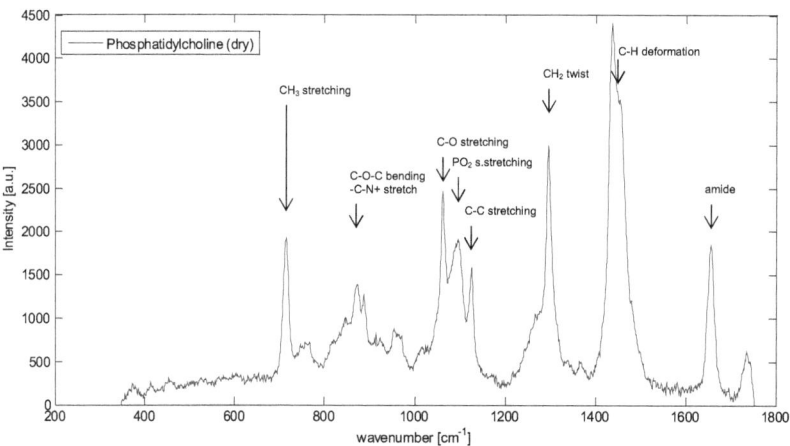

Figure 5-13 Ramam spectrum of dried phosphatidylcholine recorded with 514.5nm excitation for 10s with a grating of 1800 lines / cm

It is notable that the loading of PC_1 (Figure 5-12) also exhibits similarities to the spectrum of phosphatidylcholine (Figure 5-13). However, in the scatter plot of Figure 5-10, it is seen that this PC does not differentiate between exposure times, but more likely differentiates sampling positions with particularly high densities of lipidic bodies. Similarly, the loadings of the second PC exhibits features which may be lipidic in origin, but do not contribute to the separation of datasets.

In an attempt to simplify and elucidate this further, one can translate the data into a combined separation order, the mean distance of the data points, by describing the data points by a coordinate system along the separating dimensions. This method is commonly referred to as independent component analysis (ICA), a dimensionality reduction technique that rotates the independent dimensions into the sample space, as described in chapter 2. In ICA, the dimensions are the independent components, comparable to the loadings of the PCA, but these dimensions are not orthogonal to each other, in contrast to PCA but totally independent. Furthermore, in a necessary process for ICA, called whitening, the covariance within each spectrum is removed so each dimension samples the same independent event or source, facilitating blind source separation. Therefore, ICA can improve the separation of clusters due to its capability to separate variances and remove less relevant covariances.

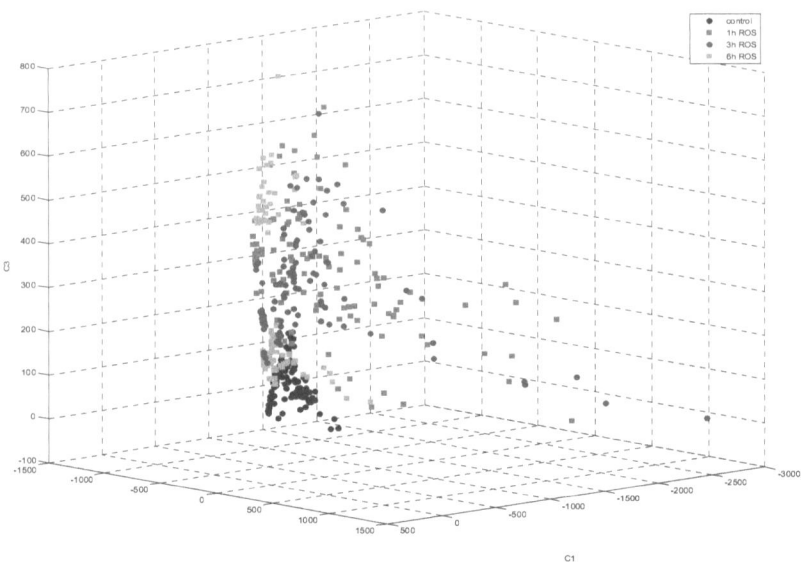

Figure 5-14 ICA scatterplot after whitening of scores on IC_1 - IC_3

The number of possible independent components is determined by the number of eigenvectors and thus, similar to PCA, the fewer components necessary to describe the majority of the captured variance, the better the model. Unfortunately, the importance of the independent components is not ordered, as is the case for PCA, and therefore it can be problematic to find the most appropriate independent components. The data that were previously fed into PCA were subsequently fed into ICA with the aim of rotating the dimensions towards the visible separation of the data in order to identify the dominating independent effect due to ROS exposure over time. In simple terms, the analysis attempts to isolate ROS induced changes in the cellular spectrum, without the other sample covariances in the identical spectral region that are caused by additional non-ROS dependent influences. The independent component scatterplot (Figure 5-14) shows a similar separation between each group observable in Figure 5-10. The only difference is that each cloud is extended horizontally and it seems that the order in the vertical direction is slightly altered. It seems that the spectra of the cells that were exposed to reactive oxygen species for 1 and 3 hours predominantly share the same subspace, whereas in the PCA scatter plot a slight separation between one and three hour exposed spectra could be observed. Thus, predominantly the 1 and 3 hour exposed samples are scattered along the horizontal axes. Notably, the samples that were exposed for 6 hours are again separated into two subgroups, one that is far away (in the vertical direction) from the control group and another which is close to the control group. Such a separation, into two groups, is indicative of a temporally defined critical recovery mechanism, the temporal anticipation of cellular damage. The separation of this dataset may prove difficult for further analysis due to the distortion of any training set for example, in partial least squares regression or even artificial neural networks (ANN), as the control group as the anchor for the targets will be distorted by the long time exposure samples that collocate in PCA and ICA. Therefore it is suggested that the samples of the 6h exposure that collocate with the controls are relabelled or excluded (e.g. selected with linear discriminant analysis (LDA) or manually).

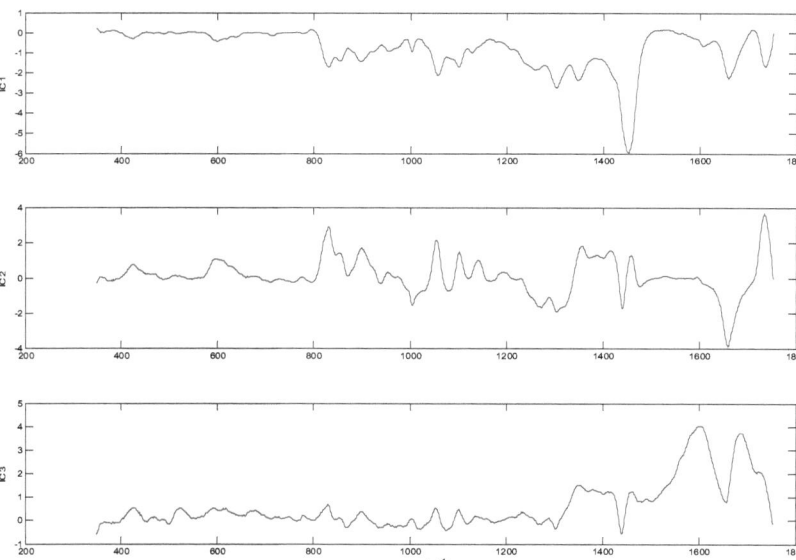

Figure 5-15 Independent component loading plot of IC_1, IC_2 and IC_3 of ROS exposed A549

The loadings plots of the three components of the independent component analysis, which are the basis for the separation or spreading of the data points in Figure 5-14, are shown in Figure 5-15. A huge similarity to the principal components one to three is observable. Similar to the PCA scatter plot, the data of the ICA scatterplot separate primarily along dimension three. In a comparison between PC_3 and IC_3, the overall similarity is marked, but small features of PC$_3$ become more prominent or even inverted as shown in Figure 5-16.

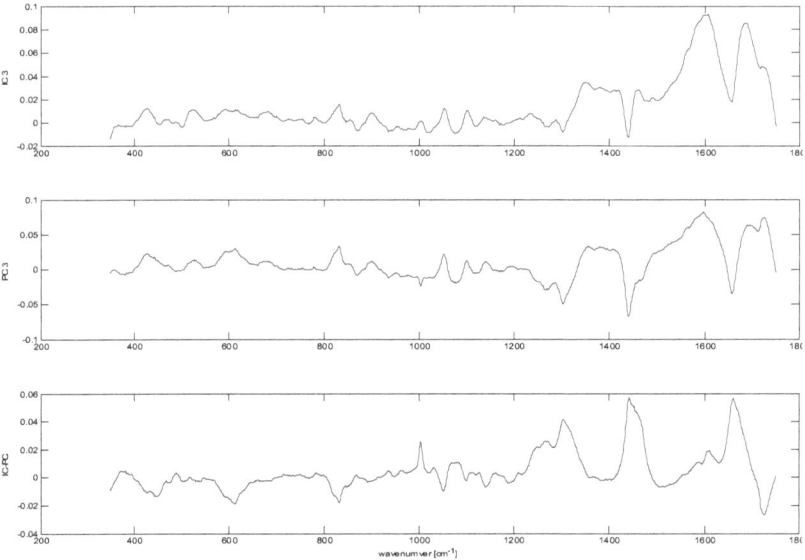

Figure 5-16 Comparison between PC$_3$ and IC$_3$

The difference loading of IC$_3$ and PC$_3$ (IC$_3$ - PC$_3$) shares several features with those of average healthy A549 cells (Figure 5-16) and proteins, whereas IC$_1$, though inverted, and IC$_2$ share several features with common phospholipids like phosphatidylcholine (Figure 5-13). Notably, however, the CH$_2$ rocking vibration at ~721 cm^{-1} of proteins and lipids does not show up as a pronounced feature in any component or spectrum. However, strong spectral indicators are given that demonstrate that the dominant cellular response to oxidative stress induced by H$_2$O$_2$ occurs in the functional groups of lipids. ROS in the form of hydrogen peroxide can turn unsaturated fatty acids into reactive oxygen species themselves (auto peroxidation), predominantly targeting the CH and CH$_2$ bonds attempting to abstract single hydrogen. However, not only lipids are exposed to the oxidative threat (lipid peroxidation), proteins and DNA/RNA are also affected. Due to the nature and design of the experiment, the externally introduced ROS predominantly interacts with the surface of the cell and the strong spectral changes are detectable across the surface (Figure 5-4). The cell membrane and its lipids and proteins seem to be the primary target of the ROS. The cellular membrane is a bi-layered structure that consists predominantly of phosphatidylcholine and a number of floating active complexes [46], comprised of three dimensionally shaped proteins. It is not surprising therefore, that most of the variation of the cellular spectra can be ascribed to features that can be found in the spectrum of phosphatidylcholine itself (Figure 5-13). Reactive oxygen species target several functional groups of this phospholipid as a key constituent of cell membranes and phosphatidylcholine has also been shown to play an active role in the decomposition of H_2O_2 [208]. Secondly, membrane proteins, especially those with metal ion functional groups, can also be damaged due to the generation of ROS in Fenton-like reaction with H_2O_2 [209].

It appears, however, that the process and effect of oxidative damage on the cellular level is reversible to a certain extent and membrane damage can be repaired. The primary membrane repair or segmentation capabilites are the probable cause of the clearly time dependent intensity response, notable already in the average spectra (Figure 5-4) and the total integrated signal intensity (Figure 5-5). Within the first three hours of exposure, the phospholipids of the cell membrane seem to be the primary targets indicated by the dispersion of the samples along IC_2. Afterwards the permanent alteration of the e.g. membrane protein and DNA may be evident as IC_1 is in some areas an inverse of IC_2. Regions within the Amide III band (~1287 cm^{-1}) seem to be more prominent in the IC_2 as well as the Amide I band at ~1656 cm^{-1} and the CH_2 band at ~1450 on the other hand is less striking. IC_2 and IC_3 share most of the features though IC_3 is of significant lower intensity and adverse in the region around Amide I. Therefore, it seems that although the membrane damage might be reversed within six hours (along IC_1), damage to the membrane proteins and ongoing repair mechanisms can be of a long term or permanent nature indicated by the split of the 6 hour exposure group along IC_3. As a final reaction to reactive oxygen species, the cell undergoes autophagic cell death, sometimes described as the very similar process of apoptosis. This kind of cell death involves the separation of cell components in membrane-enclosed vesicles, which are visible in all exposed cells (Figure 5-1). It couples the increase in vesicles with the selective autophagic degradation of peroxisomes, therefore the strong response in these membrane related features can be explained further. This membrane influence, the vesicularisation of cells as a precursor to cell death, here autophagy, could in a similar manner describe some features of the spectra from the depleted medium study, which point towards a similar increased vesicularisation (Figure 5-17). The changes in the tri-feature cluster between 1000 cm^{-1} and 1200 cm^{-1} (O-P-O backbone stretching at 1070cm^{-1}, the C-C stretching vibration at 1100cm^{-1} and the C-C skeletal stretching at 1124 cm^{-1}) observable in Figure 4-4 could be the result of similar increases in lipidic activity. In the study of Davoren et al., [75], although no internalisation of SWCNTs was observable by

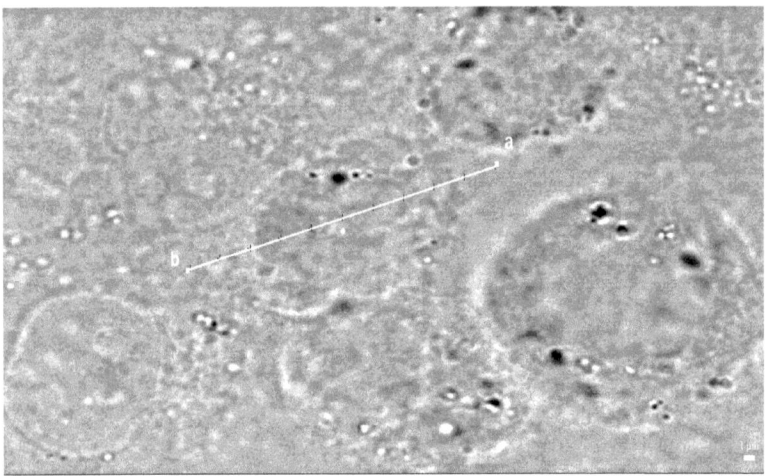

Figure 5-17 Micrograph of filter control medium exposed cells after 96 h exposure showing clear vesicularisation.

TEM, a significant change in cellular morphology was observed, associated with increases in lamellar bodies and microvilli, which was interpreted as a cellular response to nutrient deficiency and oxidative stress, as a result of medium depletion. Both aspects corroborate the above explanation of a lipid modulated response to ROS.

5.3.4 Multivariate PLS regression modeling

In an attempt to create a model for prediction of oxidative stress, the data were fed into PLS regression, taking the exposure time as the target. The preprocessing for this purpose consisted of vector normalisation, mean centring and smoothing with a Savitzky-Golay filter of second order and window size of 15 nodes. A quarter of the dataset was randomly selected as the unseen dataset and the other 75 % formed the calibration and validation dataset. In order to choose the optimum number of LV's to be retained, 'leave-one-out' cross validation was carried out on the calibration set.

Seven LV's were retained for model construction, as the RMSECV did not decrease significantly after this point. After building the model, the unseen data were presented, revealing that the prediction of exposure time did not deliver acceptable prediction precision. The maximum RMSEP 2.98 was achieved with $R^2 = 0.568$ as shown in Figure 5-18.

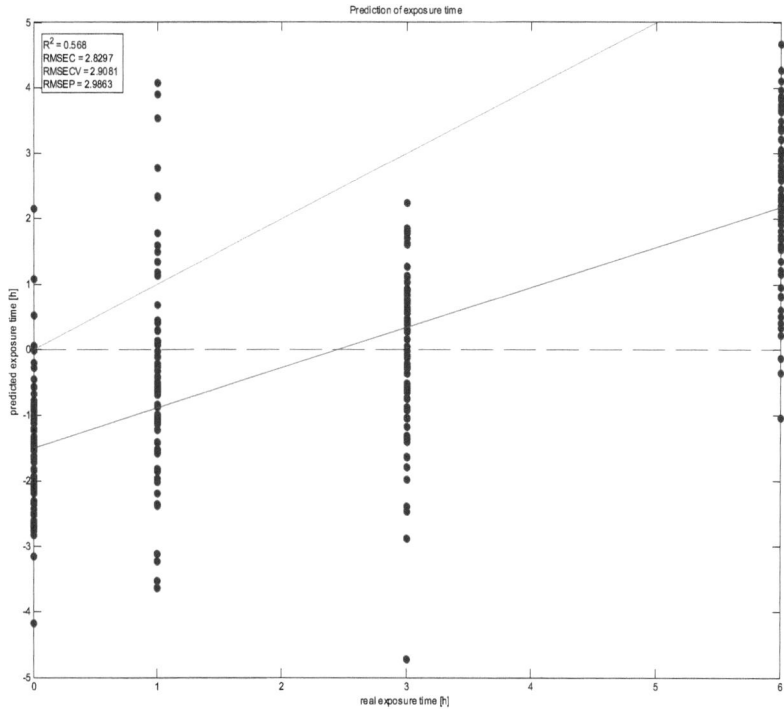

Figure 5-18 PLS regression prediction result for exposure time to ROS (full dataset)

In a second PLS regression model the data points from the six-hour exposure that co-localized with the control were removed from the dataset completely, and a second PLS regression model similar to the first one was constructed.

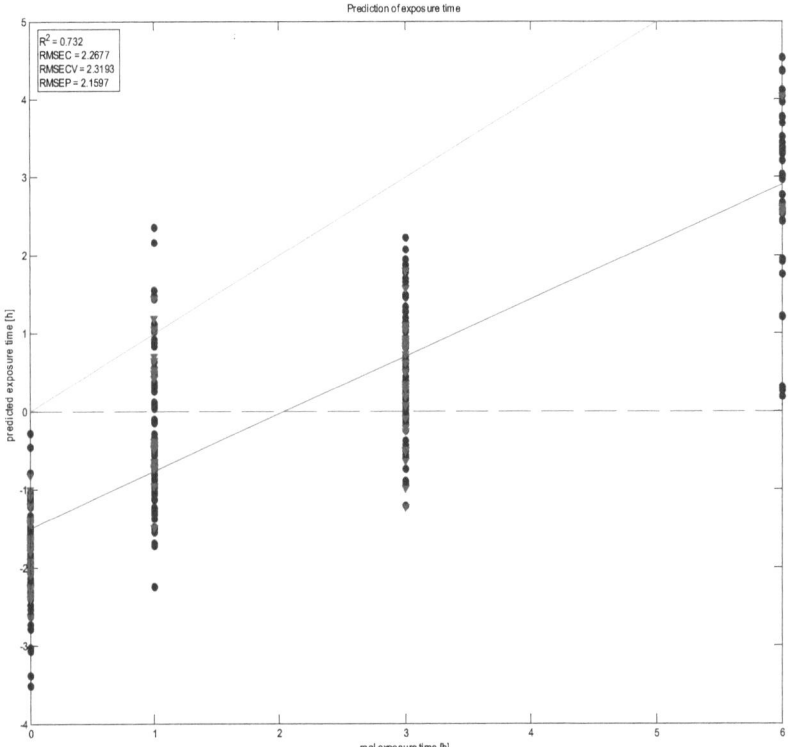

Figure 5-19 PLS regression prediction result for exposure time to ROS (corrected dataset)

By doing this, a lower RMSEP of 2.15 with a R^2 = 0.73 could be achieved which only indicates prediction for control or long-term exposure. This indicates that the targets are still problematic for a linear PLS regression model, because of the introduction of the highly nonlinear influences that are introduced via the common cellular quadratic linear responses to external influences discovered already using colorimetric assays [Aidan D. Meade, PhD Thesis, DIT, (2010)]. Therefore, it is understandable that a linear approach does only converge, and deliver a reasonable result, whilst the cellular response approaches a linear behaviour.

Based on the observations of the micrographs, Figure 5-1, which seem to indicate increased vesicle formation and the scatter plot, Figure 5-14, which seems to indicate time dependent changes, in a different approach the data were regressed against a manually created target termed 'permanent membrane damage towards autophagy', that basically is configured as a set of stages, as a precursor to autophagic cell death of just three groups [0 1 2]. For this purpose, the recorded spectra were regrouped into the group of spectra recorded from control, and those that collocate in ICA and PCA after six-hour exposure to ROS. The second group consisted of the spectra of one and three hour exposed samples, and the last group was comprised of the spectra that were exposed for six hours and did not co-localise with the control spectra. The PLS regression then was rerun with the same basic parameters in terms of cross validation and sample numbers as the previous, and much better prediction results could be generated. A predictive model with a RMSEP of 1.14 with an R^2 = 0.855 with only four latent variables could be achieved Figure 5-20.

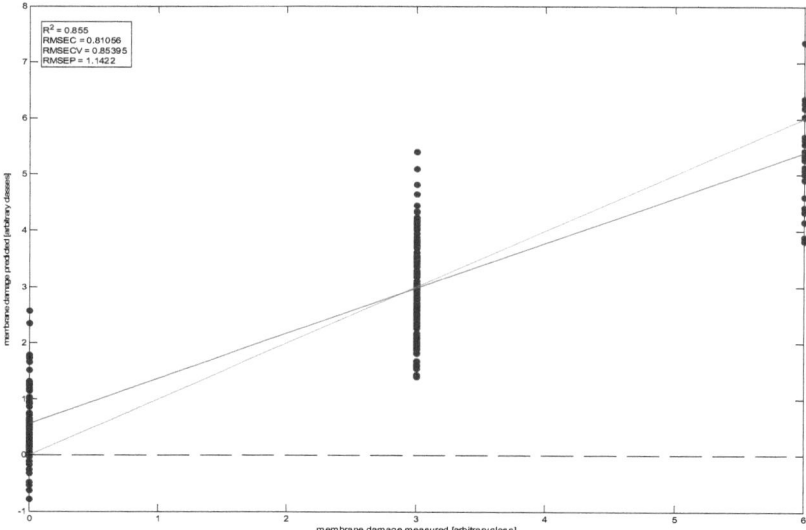

Figure 5-20 PLS regression - prediction result for predicted membrane protein damage to ROS (corrected & regrouped dataset)

Although the prediction accuracy and the model stability improved dramatically, it is still problematic to classify or differentiate each group correctly. It seems that the best differentiation can be made between the intact and the defect membrane or higher number of sub membrane vesicles, as a consequence of compartmentation of the defect membrane fragments or peroxidised phospholipids. The centre group could probably be described as those recordings where membrane repair is still ongoing either towards an intact, cell membrane without an increased number of vesicles, or towards autophagic programmed cell death [47]. Generally, it should be considered that nonlinear PLS regression in future might improve the construction of a predictive model.

5.4 Discussion and Conclusions

This chapter explored the strong influence of reactive oxygen species on the physiology of A549 lung cells. Oxidative stress has been identified as a primary mechanism of the toxic response of cells to nano-particulate exposure. The cellular changes associated with oxidative stress can be monitored with Raman spectroscopy. Single and multi variate analysis indicate a broad range of effects at a molecular level. A primary effect of damage to the phospholipid bilayer of the cell membrane, which is at least partially reversible, is suggested. A secondary effect of non-reversible protein damage is also suggested. Reactive oxygen species cannot be understood as toxins in themselves, however and the cellular responses cannot be monitored in terms of common spectroscopic markers of acute toxicity of chemical species as shown in Figure 5-9.

Multivariate analyses corroborate the assumption that predominantly membrane damage is the mode of action. In the scatter plots, the samples that were exposed for one and three hours co-occupy the same space, though they are spread along the first and second dimension indicating that between one and three hours, similar biochemical courses of events take place. Along the third dimension, the data separate differently, with a clear split of the six hour exposed ones, indicating that after six hours, the initial influence of reactive oxygen species is reversed. Dimension three in either PCA or ICA, are very similar and indicate subtle changes over the complete spectral range (Figure 5-16). As independent components can be understood as covariance free, the difference between both can be understood as a blurring function as described in section 2.4.8.5.1. between PCA and ICA. Not surprisingly, this difference, among others, shares similarities with the average spectrum of a healthy A549 cell. As in this study mainly spectra of the cell membrane and the underlying cell organelles are taken, independent component three therefore, as the dimension of separation, becomes the most important and exhibits mainly changes in protein and carbohydrate attributed spectral regions.

In the attempt to model the data via PLS regression, it becomes obvious that the inclusion of the full dataset makes it impossible to achieve an accurate prediction. Only by altering the targets towards three classes of arbitrary character, for the ongoing membrane damage and repair, can improvement be made. This again demonstrates the huge variability of cellular spectra, but still explains what is going on in the cell membrane exposed to oxidative threat.

The strong responsiveness of the A549 cells to oxidative stress was demonstrated and a timeframe for this response was outlined. Overall, the capability of Raman spectroscopy to detect and identify biological changes in the cell membranes induced by oxidative stress was demonstrated and an additional building block for the understanding of the toxicity of single walled carbon nanotubes was established.

Chapter 6 : Discussion and conclusion

6.1 Introduction

In this chapter, the results and conclusions of the previously described work will be put into context and the relationship between each single experiment and the overall framework will be established. The interplay of the three major components of this study, the nano particle induced toxicity, Raman spectroscopy and cellular responses will be elucidated. The direct and indirect toxicological aspects of nano particle exposure will be discussed using the example of single walled carbon nanotubes inducing a toxic threat to alveolar cells *in vitro* as a model of occupational exposure by respiration. The diverse range of cellular responses to a variety of influencing threats, monitored by Raman spectroscopy, enhanced with adapted preprocessing methods and holistic analytical approaches will be demonstrated. Finally, it will be concluded that Raman spectroscopy can outperform common colorimetric approaches to cellular toxicity, although the need to revisit key aspects of the broad application of Raman spectroscopy in terms of signal processing and the overall explanatory power of this technology and experimental design is highlighted.

6.2 Nano-toxicity and Spectroscopy

Evaluation of the toxicology of nano materials is of critical importance, as materials that might appear non-toxic on the micro scale can change their toxicological behaviour on the nano scale [2, 64]. In assessing the toxicology of nanoparticles, common colorimetric dyes have proven to be problematic due to their ability to interact with the employed toxin [171]. Furthermore, *in vitro* colorimetric cytotoxicity assays capture only partial aspects of the cellular reaction to the threat of the toxic agent. Although the clonogenic assay was introduced as a holistic approach to assess the toxicity of nano particles [99], it is a phenomenological rather than mechanistic assay. The toxic effect can be a combination of direct and indirect influences [101, 170] and even worse, can be dependent on the metabolic or cell cycle stage [92, 210]. The assay sheds little light on the mechanism of interaction with the cells or the process of cellular response. As an alternative technology, spectroscopy seems to be a valid method to assess the cellular changes as a consequence of exposure to toxins [170].

Biospectroscopy as such is an exciting field and from the beginning of this work in November 2006 until today, the number of scientific articles dealing with Raman spectroscopy for analysis of biological systems alone, that are listed in Pub MED, doubled. Spectroscopic analysis can potentially capture the total chemical composition of the sample in the measured spot or area [211]. Therefore, it offers the same holistic result as the cytotoxicity assays with the advantage of molecular scale information which may lead to further understanding of the mechanism of response. However, spectroscopic responses are very rich in detail. They are prone to the influence of the necessary spectral preprocessing in order to extract features of value [122]. Generally, it is very important to understand the source of any influences that obscure the actual meaningful signal [123, 212]. For example, in infrared spectroscopy, only by truly understanding the origin of the so-called dispersion artefact can it accurately be removed [213-215]. Raman spectroscopy suffers from similar influences. Although it has proven to be a valuable tool in material sciences, Raman spectra are likely to be obscured by strong features such as the excitation source that may be outside the spectral window [133]. The strong background often observed for tissue samples is frequently interpreted as sample fluorescence, but an origin in diffuse scattering is more likely [216]. Raman has the advantage that it can be applied *in situ* [217], with an increasing number of miniaturised systems on the market [218], but a significant amount of data preprocessing is usually required. The potential danger of preprocessing in general is the capability to introduce artefacts into the data which can lead to misinterpretation of experimental results, thus the benefits of optimising the signal to noise from the outset cannot be overstated.

6.3 Improved Methods

In chapter 2 the various influences of preprocessing and postprocessing were illustrated and discussed [122]. A whole battery of preprocessing tools was introduced, each designed to tackle individual aspects of the recorded spectra. The majority of the preprocessing methods introduced are commonly used in differing combinations in the field of spectroscopy [219, 220]. The preprocessing methods were tested regarding their ability to improve spectral data without introducing additional features that might influence the explanatory power of the spectra [219]. While sometimes the effects are subtle, they can at other times be very strong. In assessing work in the field, it is crucial to know that the described effects are not the result of artefacts introduced by preprocessing. In Raman spectroscopy, the intensity and the frequency of a feature are employed to determine the chemical composition [123] or changes to it as a result of physical, chemical or biological processes. In a number of reference tables, known features are listed with similar but different bandwidths [93, 105, 221], sometimes even at different positions, without mentioning which preprocessing methods led to the assignment. Therefore, it is difficult to properly and with precision, pinpoint the true change in a spectrum. Furthermore, effects are described that do not influence the spectra linearly [219]. It becomes clear that sharp features can be easily shifted by underlying or neighbouring broad features. Broad features can comprise of wanted and unwanted contributions. Thus, depending on the background or baseline removal that was applied, feature shifts can be introduced by preprocessing, which can render any information derived from spectral shifts somewhat questionable in the application of Raman spectroscopy for biological investigations. It is therefore suggested that the preprocessing should be limited, meticulously documented and disclosed entirely. It is suggested that any removal of baseline or substrate [117, 121] should be performed in a noise free manner in order not to induce further variance of features. The processing methods employed target the spectral signal quality, endeavouring to employ the complete knowledge of the experimental setup. Therefore, it is important to record the signals with the maximum signal to noise ratio available, while maintaining sample integrity. Data analysis methods employed obviously suffer in various ways from the variance that is induced by the pre and post-processing. Residual baseline not only results in altered intensity of spectra, but also the broadening of features in multivariate analysis. Excessive noise filtering can remove features completely. The simple subtraction of background can induce additional noise, inverse substrate features and other unwanted effects. Ultimately, it is desirable to optimise the signals themselves to minimise the need for preprocessing. Therefore, it becomes clear that overinterpretation as a result of either the preprocessing or the analysis has to be minimised.

6.4 Implications of the direct exposure

The direct exposure experiment presented in chapter 3, in which lung epithelial cells are exposed to single walled carbon nanotubes directly, establishes Raman spectroscopy as a valid technology to assess toxicity. Measured effects can be correlated directly with toxicological endpoints derived from clonogenic assays, and previously identified markers for direct toxicity [99, 170]. Although the results of this study indicate that Raman spectroscopy can pick up spectral changes that are relatable to cytotoxic markers and clonogenic endpoints, it is not clear whether the induced spectral changes as markers of toxicity are of primary or secondary character, as previous studies suggest that the mode of action could be, among others, a consequence of starvation and/or the direct exposure to nanotubes [101]. The overall experimental design, wich led to the results shown, mimics the real world exposure scenario of nano material dust inhalation and indicates the complexity of the toxicity of nano particles. Not only do the nano particles in themselves elicit a complex response, but it is clear that the ill defined dispersion leads to an unpredictable actual exposure, adding another dimension of variance. The cells themselves and the data processing probably added their share of variance to the spectroscopic measurements. Thus, although the results as such look very promising, the influence of aspects of sample dispersion and extensive data filtering, pre and post processing, should be assessed.

Consequently, the influences of data preprocessing were targeted in the subsequent studies, by separating the direct presence of single wall carbon nanotubes and their known influences from the exposed cells. Though the experiment mimicked the real world exposure, and the results partially match the expectations, on the one hand, Raman spectroscopy was demonstrated as a capable technology to access toxicity of SWCNT. On the other hand, it was shown that especially the use of single wall carbon nanotubes, to mimic occupational risks of NP exposure, strongly influence the necessary experimental design.

6.5 Consequences of indirect exposure

In the subsequent study, the potential influence of ill defined dispersion of and cellular exposure to SWCNTs was minimised by medium filtering prior to exposure, mimicking the indirect exposure study of Casey et al. [101]. In the spectroscopic study, no gelatin coating was employed to minimise any potential variance due to inhomogeneous coating of the substrate. In terms of data preprocessing only the noise free rubberband baseline and the modelled substrate was removed which made no filtering necessary. Chapter 4, displays the results of the exposure of A549 lung cells to medium that was nutrient depleted by various doses of single wall carbon nanotubes, as it was suggested from previous studies that depletion as a secondary toxic effect can be the primary mode of action when cells are exposed to single wall carbon nanotubes [101]. Therefore, it was attempted to model this starvation as a single effect. With univariate analysis and the previously employed markers for toxicity, no clear systematic dose dependence could be established. In all cases, substantial variation in the cellular spectral response was observable, and thus subtle changes could not be identified. It could be shown, however, that the filtering of the SWCNT exposed medium systematically reduced the protein content of the culture medium, as shown by the concentration dependent reduction of the intensity of the Amide I band in figure 4-3. Thus the dispersion difficulties that emerged in the direct exposure study of chapter 3 should be eliminated or minimised. However, the process of adsorption of components from the medium and indeed the filtration process itself may entail more subtle effects which affect the composition of the medium and thus the impact on the cells after chronic exposure.

Multivariate analysis showed similar results for all exposed spectra, indicating no direct depletion response. The multivariate analysis of control medium and filter control medium exposed cells show differences in addition to those of the SWCNT depleted medium exposed cells. The cellular response as a result of starvation is suggested to be not only caused by the medium depletion by SWNT exposure, but also by the long exposure time. It is suggested that all cell samples show signs of reduced health, although the control, incubated in complete medium, suffered less from the effects.

In the clonogenic assay, such long exposure times are required for colony formation, and the clonogenic study was employed as colorimetric assays were deemed to be in accurate due to the interaction of the dyes with the SWNTs. Overall, however, the exposure time of the initial study of Casey et al. ref no. was appraised as being inappropriately long for an accurate assessment of the spectroscopic response. This gives rise to consideration of the overall experimental design.

The absence of a systematic dose dependence of the cellular spectra may suggest that the cells as such have various ways of action to evade starvation [46]. The similarity between all exposed cells may indicate that the cells are in a similar state of health after 96 h which might be due to cell synchronisation, as A549 can be synchronized by serum withdrawal, and in this case the experiment might only show that all samples are at a similar stage prior to cell death induced by starvation [222].

As compared to colorimetric assays commonly employed to analyse the effects of medium depletion by SWCNT, however, it appears that Raman spectroscopy can identify responses at lower concentrations.

Retrospectively, it could be argued that the experimental design was too simplistic to capture all varieties of modes of action and cellular response. Therefore, it seems that the question " What happens to A549 cells when they are starved? " is too vague to be answered by Raman spectroscopy, although the observed results share similarities to a parallel study employing colorimetric assays as a viability test [101].

6.6 Implications of hydrogen peroxide exposure

In a further approach to elucidate another specific toxic effect of single wall carbon nanotubes, in chapter 5 the same cell type as in the previous studies was exposed to an immuno cytokine [46], the common toxic mediator hydrogen peroxide. This agent is involved in cell signalling and oxidative stress, but also known to be involved in nano particle toxicity [43, 66]. Exposure to this agent resulted in significant and clearly observable spectral responses. The spectroscopic data suggest that the targeted structures of the cell are the membranes [208, 223], either involved in peroxidation of lipids or the generation of vesicles. The importance of lipids in this process is obvious, as unsaturated fatty acids are prone to peroxidation [46] a process which in itself induces oxidative stress, and the fact that multiple unsaturated fatty acids can act as scavengers of ROS [46]. It seems that changes to the contributions of phospholipids are the main spectroscopic fingerprint of hydrogen peroxide induced oxidative stress to the cell. The A549 lung cell in itself seems likely to respond to H_2O_2. It produces a phosphatidylcholine (PC) based surfactant , namely DPPC, in lamellar bodies to clear particular surface threats [75]. The presence of this surfactant and SWCNT induces drastic changes in the generation of ROS of exposed A549 [45]. Secondly, the time dependence of this oxidative threat could be visualised. It is suggested that the cells respond to the exposure to hydrogen peroxide in a time dependent fashion, possibly indicating the presence of permanent damage or the inevitability of pending autophagic cell death [47]. Although hydrogen peroxide is a relatively unreactive oxygen species, it might be understood that the cells are suffering from Fenton-like reactions that turn H_2O_2 into a internal ROS threat [45]. Additionally as hydrogen peroxide is the first messenger of an immune response, secreted by neutrophils, the lung cells themselves could prepare for the release of DPPC responding to the extracellular H_2O_2 as a sign of external threat, requiring additional surfactant to sustain the lung epithelium. Remarkably, a significant proportion of the cell population recover to the control state after 6 h.

Although the spectroscopic changes indicate a systematic evolution of phospholipids in the cellular response to oxidative stress, it seems to be difficult to create a multivariate model to describe the response. PLS regression are linear models however, and thus choice of target is critical to the accuracy of the model. Unfortunately, the kernels of nonlinear PLS regression models are not easily accessible such that an arbitrarily nonlinear dependence of spectral response on a target response or external dose can be more accurately modelled. A model still could be generated in an alternative approach exploiting the permanent damage to the cell membrane, which might be present after overcoming the direct oxidative stress.

Nevertheless, the study shows that the response of A549 cells to an externally applied oxidative stress as well as the time evolution of the response can be identified and characterised by Raman spectroscopy, rendering this technology as a valid tool to detect oxidative stress.

6.7 Conclusion and future aspects

As a result of this study, the selective applicability of Raman spectroscopy, for the analysis of cellular toxicity, is clear. This work clearly displays that nanoparticles can interact with the cell in various ways and that the cell itself can respond in similarly varied manner. The toxicity of carbon nanotubes can be quantified, on the basis of a holistic approach by Raman spectroscopy. More specific mechanistic aspects of the exposure to nano particles can be detected and modelled, as long as the cellular response is not diverse.

The overall conclusion therefore is that the validity of Raman spectroscopy for the analysis of toxicity is established. Much work is required however to optimise sample responses and minimise effects of data processing. As a model system, carbon nanotubes are notoriously difficult to disperse in aqueous media and efficient dispersion can only be achieved by significant processing (sonication, chemical functionalisation) or through use of a dispersion agent [188]. Efficient dispersion is highly concentration dependent, and thus dose dependence would be effectively of a material whose properties vary across a range. In the studies of Davoren et al., Herzog et al., and Casey et al.,[45, 75, 99, 101, 102] minimal dispersion was employed to mimic occupational exposure by dust inhalation, and this study was designed to build on these studies to explore the potential of Raman spectroscopy. In future work, some aspects should be revisited, as it would be beneficial to have backward comparable results by updated processing techniques and their consequent disclosure. In future experiments a better defined material, with a known systemic toxic response and mechanism could serve as an appropriate model system to establish the potential of Raman spectroscopy in this field. Such a model system could explore variations in size and surface chemistry, as it is available for example in functionalised quantum dots [224-227]. With the developed methods and the immense progress in signal processing the true value of Raman spectroscopy might now even more serve as a viable tool to investigate intact living cells or its fractions on the exposure to a broader variety of stimuli, capturing the total chemical response. Therefore it can be well integrated into the field of Systems Biology as it adds complimentary, colocational and purely chemical, information as an additional dimension to it.

Abbreviations and acronyms

A549	adenocarcinoma human alveolar basal epithelial cells
ATCC	american type culture collection
BSS	Blind source separation
CC	confidence curve
CCD	Charge coupled device
CNT	carbon nano tube
D band	disorder induced band
DMEM	Dulbecco's modified eagle medium
DNA	desoxy ribonucleic acid
DRMS	dynamics assisted moving median filter
EU	European union
EMSC	extended multiplicative scatter correction
FBS	foetal bovine serum
FIR	finite impulse response
G band	tangential vibrational mode (band)
GA	genetic algorithm
HCA	Hierarchical cluster analysis
HiPco	high-pressure CO conversion
HIS	Hyper spectral imaging
ICA	Independent component analysis
IR	Infra red
LV	latent variable
MA	Moving average
MATLAB	matrix laboratory
MIPS	million instructions per second
MRI	magnet resonance imaging
MTT	mitochondrial tetrazole assay
MVA	Multivariate analysis
MWCNT	multi walled carbon nano tube
NIST	national institute of standards
NP	Nano particle
PBS	Phosphate buffered saline
PC	principal component
PCA	principal component analysis
PLS	partial least squares regression

PMC	Pub MED Central
RMS	moving median filter
RMSE	root mean squared error
RMSEC	root mean squared error of calibration
RMSECV	root mean squared error of cross validation
RMSEP	root mean squared error of prediction
ROI	region of interest
ROS	reactive oxygen species
S/G	Savitzky & Golay
SNR	signal to noise ratio
SRM	standard reference material
SWCNT	single walled carbon nano tube
TEM	transmission electron microscopy
TPa	tera pascal
UV	ultra violet
WST	water soluble tetrazole assay

References

1. Oberdörster, G., et al., *Principles for characterizing the potential human health effects from exposure to nanomaterials: elements of a screening strategy.* Part Fibre Toxicol, 2005. 2: p. 8.
2. Dowling, A.P., et al., *Nanoscience and nanotechnologies: opportunities and uncertainties* in *The Royal Society*. 2004. p. 11.
3. Chou, S.G., et al., *Optical characterization of DNA-wrapped carbon nanotube hybrids.* Chemical Physics Letters, 2004. 397(4-6): p. 296-301.
4. Kang, D., D. Kim, and E. Sim, *Size-dependent quantum dynamical influence of metal nanoparticles on surface plasmon resonance* Proceedings of SPIE, the International Society for Optical Engineering, 2007. 6479.
5. Salata, O.V., *Applications of nanoparticles in biology and medicine.* Journal of Nanobiotechnology, 2004. 2(1): p. 3.
6. Ketelson, H.A., N.D. McQueen, and *Use of inorganic nanoparticles to stabilize hydrogen peroxide solutions*, U. Patent, Editor. 2008, Alcon, Inc.
7. Fujishima, A., T.N. Rao, and D.A. Tryk, *TiO2 photocatalysts and diamond electrodes*, in *Electrochimica acta*. 2000, Elsevier, Oxford, Royaume-Uni Gerischer Symposium No1, Berlin, Allemagne (23/06/1999)
8. Tenne, R., *Inorganic nanotubes and fullerene-like nanoparticles.* Nat Nano, 2006. 1(2): p. 103-111.
9. Feng, S.-S., *Quantum dots fluorescent protein pairs developed as novel fluorescence resonance energy transfer probes.* 2008. p. 279-281.
10. Nam, J.M., C.C. Thaxton, and C.A. Mirkin, *Nanoparticles-based bio-bar codes for the ultrasensitive detection of proteins.* 2003. p. 1884 - 1886.
11. Geho, D.H., et al., *Nanoparticles: potential biomarker harvesters.* Curr Opin Chem Biol, 2006. 10(1): p. 56-61.
12. Iijima, S., *Helical Microtubules of Graphitic Carbon.* Nature, 1991. 354(6348): p. 56-58.
13. Gregan, E., et al. *Stokes/anti-Stokes Raman spectroscopy of high-pressure carbon oxide (HiPco) single-walled carbon nanotubes.* in *Opto-Ireland 2002: Optics and Photonics Technologies and Applications.* 2003. Galway, Ireland: SPIE.
14. Zheng, L.X., et al., *Ultralong single-wall carbon nanotubes.* Nat Mater, 2004. 3(10): p. 673-676.
15. Dekker, C., *Carbon nanotubes as molecular quantum wires.* Physics Today, 1999. 52: p. 22-28.
16. Gokcen, T., C.E. Dateo, and M. Meyyappan, *Modeling of the HiPco process for carbon nanotube production. II. Reactor-scale analysis.* J Nanosci Nanotechnol, 2002. 2(5): p. 535-44.
17. T. Guo, et al., *Catalytic growth of single-walled manotubes by laser vaporization.* Chemical Physics Letters, 1995. Volume 243(Issues 1-2): p. 49-54.
18. Colomer, J.-F., et al., *Synthesis of Single-Wall Carbon Nanotubes by Catalytic Decomposition of Hydrocarbons* Chemical Communications, 1999(14): p. 1343-1344
19. Wagner, H.D., *Reinforcement*, in *Encyclopedia of Polymer Science and Technology*, J.I. Kroschwitz, Editor. 2003, Wiley-Interscience: New York. p. 94-115.

20. Sinnott, S.B. and R. Andrews, *Carbon Nanotubes: Synthesis, Properties, and Applications.* Critical Reviews in Solid State and Material Sciences, 2001. 26: p. 145-249.
21. Li, Y., et al. *The Latent Toxic Effects of Carbon Nanotube Serving as Biomedicine.* in *Bioinformatics and Biomedical Engineering, 2007. ICBBE 2007. The 1st International Conference on.* 2007.
22. Girifalco, L.A., M. Hodak, and R.S. Lee, *Carbon nanotubes, buckyballs, ropes, and a universal graphitic potential.* Physical Review B, 2000. 62(19): p. 13104.
23. Wang, H. and E.K. Hobbie, *Amphiphobic Carbon Nanotubes as Macroemulsion Surfactants.* 2003. p. 3091-3093.
24. Kataura, H., et al. *Bundle effects of single-wall carbon nanotubes.* in *American Institute of Physics Conference Series.* 2000.
25. Srivastava, D., C. Wei, and K. Cho, *Nanomechanics of carbon nanotubes and composites.* Applied Mechanics Reviews, 2003. 56(2): p. 215-230.
26. Dresselhaus, M.S., G. Dresselhaus, and A. Jorio, *Unusual Properties and Sstructure of Carbon Nanotubes.* 2004. p. 247-278.
27. Meo, M. and M. Rossi, *Prediction of Young's modulus of single wall carbon nanotubes by molecular-mechanics based finite element modelling.* Composites Science and Technology, 2006. 66(11-12): p. 1597-1605.
28. del Valle, M., et al., *Tuning the conductance of a molecular switch.* Nat Nano, 2007. 2(3): p. 176-179.
29. Cleuziou, J.P., et al., *Carbon nanotube superconducting quantum interference device.* Nat Nano, 2006. 1(1): p. 53-59.
30. Urszula Dettlaff-Weglikowska, M.K.B.H.V.S.J.W.J.L.S.R., *Conducting and transparent SWNT/polymer composites.* 2006. p. 3440-3444.
31. Tran, T.D., et al., *Clinical applications of perfluorocarbon nanoparticles for molecular imaging and targeted therapeutics.* Int J Nanomedicine, 2007. 2(4): p. 515-26.
32. Lanza, G.M., et al., *Magnetic resonance molecular imaging with nanoparticles.* J Nucl Cardiol, 2004. 11(6): p. 733-43.
33. Stampfer, C., A. Jungen, and C. Hierold. *Fabrication of discrete carbon nanotube based nano-scaled force sensors.* in *Sensors, 2004. Proceedings of IEEE.* 2004.
34. Heller, D.A., et al., *Optical detection of DNA conformational polymorphism on single-walled carbon nanotubes.* Science, 2006. 311(5760): p. 508-11.
35. Zanello, L.P., et al., *Bone Cell Proliferation on Carbon Nanotubes.* 2006. p. 562-567.
36. Mazzatenta, A., et al., *Interfacing Neurons with Carbon Nanotubes: Electrical Signal Transfer and Synaptic Stimulation in Cultured Brain Circuits.* 2007. p. 6931-6936.
37. Xue, Y. and M. Chen, *Dynamics of molecules translocating through carbon nanotubes as nanofluidic channels.* Nanotechnology, 2006. 17(20): p. 5216-5223.
38. Gogoci, U.G., *Nanotubes and Nanofibers*, ed. Y. Gogotsi. Vol. 1. 2006, New York: CRC Press, Taylor & Francis. 248.
39. Dowling, A.P., *Development of nano technologies.* Nanotoday, 2004: p. 6.
40. Muller J., Huaux F., and Lison D., *Respiratory Toxicity of multiwalled carbon nanotubes: How worried should we be ?* Carbon, 2006. 44(6): p. 1048-1056.

41. Franco, A., et al., *Limits and prospects of the "incremental approach" and the European legislation on the management of risks related to nanomaterials.* Regulatory Toxicology and Pharmacology, 2007. 48(2): p. 171-183.
42. Borm, P., et al., *The potential risks of nanomaterials: a review carried out for ECETOC.* 2006. p. 11.
43. Nel, A.E., et al., *Understanding biophysicochemical interactions at the nano-bio interface.* Nat Mater, 2009. 8(7): p. 543-57.
44. Pryor, W.A., *Oxy-radicals and related species: their formation, lifetimes, and reactions.* Annu Rev Physiol, 1986. 48: p. 657-67.
45. Herzog, E., et al., *Oxidative stress response of human lung epithelial cells upon carbon nanoparticle exposure depends on dispersion medium* - Toxicology and Applied Pharmacology, 2008.
46. Löffler, G. and P.E. Petrides, *Biochemie und Pathobiochemie.* 5. ed. Vol. 1. 1975, Berlin Heidelberg New York: Springer Verlag. 1155.
47. Yu, L., et al., *Autophagic programmed cell death by selective catalase degradation.* Proceedings of the National Academy of Sciences of the United States of America, 2006. 103(13): p. 4952-4957.
48. Marquis, B.J., et al., *Analytical methods to assess nanoparticle toxicity.* The Analyst, 2009. 134(3): p. 425-39.
49. Sayes, C.M., et al., *Nano-C60 cytotoxicity is due to lipid peroxidation.* Biomaterials, 2005. 26(36): p. 7587-7595.
50. Shvedova, A.A., et al., *Vitamin E deficiency enhances pulmonary inflammatory response and oxidative stress induced by single-walled carbon nanotubes in C57BL/6 mice.* Toxicology and Applied Pharmacology, 2007. 221(3): p. 339-348.
51. Kode, A., S.-R. Yang, and I. Rahman, *Differential effects of cigarette smoke on oxidative stress and proinflammatory cytokine release in primary human airway epithelial cells and in a variety of transformed alveolar epithelial cells.* 2006. p. 132.
52. Casey, A., et al., *Single walled carbon nanotubes induce indirect cytotoxicity by mediumdepletion in A549 lungcells.* Toxicology Letters, 2008.
53. Amiji, M.M., T.K. Vyas, and L.K. Shah, *Role of Nanotechnology in HIV/AIDS Treatment: Potential to Overcome the Viral Reservoir Challenge.* Discov Med, 2006. 6(34): p. 157-62.
54. Kalbacova, M., et al., *The study of the interaction of human mesenchymal stem cells and monocytes/macrophages with single-walled carbon nanotube films.* 2006. p. 3514-3518.
55. Stone, V. and K. Donaldson, *Nanotoxicology: Signs of stress.* Nat Nano, 2006. 1(1): p. 23-24.
56. Kang, B., et al., *Intracellular uptake, trafficking and subcellular distribution of folate conjugated single walled carbon nanotubes within living cells.* Nanotechnology, 2008. 19(37): p. 375103.
57. Nicola, M.D., et al., *Effect of different carbon nanotubes on cell viability and proliferation.* Journal of Physics: Condensed Matter, 2007. 19(39): p. 395013.
58. Kagan, V.E., H. Bayir, and A.A. Shvedova, *Nanomedicine and nanotoxicology: two sides of the same coin.* Nanomedicine : the official journal of the American Academy of Nanomedicine, 2005. 1(4): p. 313-316.
59. Kandlikar, M., et al., *Health risk assessment for nanoparticles: A case for using expert judgment*, in *Nanotechnology and Occupational Health*. 2007. p. 137-156.

60. Liu, C.H., et al., *Airway mechanics, gas exchange, and blood flow in a nonlinear model of the normal human lung.* J Applied Physiol, 1998. 84(4): p. 1447-69.
61. Shvedova, A.A., et al., *Unusual inflammatory and fibrogenic pulmonary responses to single-walled carbon nanotubes in mice.* American J Physiol Lung Cell Mol Physiol, 2005. 289(5): p. L698-708.
62. Lacerda, L., et al., *Carbon nanotubes as nanomedicines: from toxicology to pharmacology.* Adv Drug Deliv Rev, 2006. 58(14): p. 1460-70.
63. Bender, A.R., et al., *Efficiency of nanoparticles as a carrier system for antiviral agents in human immunodeficiency virus-infected human monocytes/macrophages in vitro.* Antimicrob Agents Chemother, 1996. 40(6): p. 1467-71.
64. Helland, A., *Nanoparticles: A Closer Look at the Risks to Human Health and the Environment,* in *the international institute for industrial environmental economics.* 2004, Lund University: Sweden. p. 105.
65. Pacurari, M., et al., *Raw single-wall carbon nanotubes induce oxidative stress and activate MAPKs, AP-1, NF-kappaB, and Akt in normal and malignant human mesothelial cells.* Environ Health Perspect, 2008. 116(9): p. 1211-7.
66. Oberdörster, E., et al., *Ecotoxicology of carbon-based engineered nanoparticles: Effects of fullerene (C60) on aquatic organisms.* Carbon, 2006. 44(6): p. 1112-1120.
67. Liu, A., et al., *Toxicological effects of multi-wall carbon nanotubes in rats.* Journal of Nanoparticle Research, 2008. 10(8): p. 1303-1307.
68. Dong, L., et al., *Cytotoxicity of single-walled carbon nanotubes suspended in various surfactants.* Nanotechnology, 2008. 19(25): p. 255702.
69. Muller J., H.F., Moreau N., Misson P., Heilier J-F., Delos M., Arras M., Fonseca A., Nagy J. B., Lison D.,, *Respiratory Toxicity of multiwalled carbon nanotubes.* Toxicology and applied Pharmacology, 2005(207): p. 221-231.
70. Donaldson, K., et al., *Concordance between in vitro and in vivo dosimetry in the proinflammatory effects of low-toxicity, low-solubility particles: the key role of the proximal alveolar region.* Inhal Toxicol, 2008. 20(1): p. 53-62.
71. Garza, K.M., K.F. Soto, and L.E. Murr, *Cytotoxicity and reactive oxygen species generation from aggregated carbon and carbonaceous nanoparticulate materials.* International Journal of Nanomedicine, 2008. 3(1): p. 83-94.
72. Manna, S.K., et al., *Single-walled carbon nanotube induces oxidative stress and activates nuclear transcription factor-kappaB in human keratinocytes.* Nano Lett, 2005. 5(9): p. 1676-84.
73. Geys, J., et al., *In vitro study of the pulmonary translocation of nanoparticles: a preliminary study.* Toxicol Lett, 2006. 160(3): p. 218-26.
74. Pulskamp, K., S. Diabate, and H.F. Krug, *Carbon nanotubes show no sign of acute toxicity but induce intracellular reactive oxygen species in dependence on contaminants.* Toxicol Lett, 2007. 168(1): p. 58-74.
75. Davoren, M., et al., *In vitro toxicity evaluation of single walled carbon nanotubes on human A549 lung cells.* Toxicol In Vitro, 2007. 21(3): p. 438-48.
76. Casey, A., et al., *Spectroscopic analysis confirms the interactions between single walled carbon nanotubes and various deyes commonly used to assess cytotoxicity.* Carbon, 2007. 1(4430).
77. Herzog, E., et al., *A new approach to the toxicity testing of carbon based nanomaterials – the clonogenic assay.* Toxicology Letters, 2007.

78. Plumb, J.A., *Cell Sensitivity Assays: Clonogenic Assay*, in *Cancer Cell Culture*. 2004. p. 159-164.
79. Franken, N.A.P., et al., *Clonogenic assay of cells in vitro*. Nat. Protocols, 2006. 1(5): p. 2315-2319.
80. Monteiro-Riviere, N.A. and A.O. Inman, *Challenges for assessing carbon nanomaterial toxicity to the skin*. 2006. p. 1070 - 1078.
81. Ozturk, S., et al., *Adaptation of cell lines to serum-free culture medium*. Hybrid Hybridomics, 2003. 22(4): p. 267-72.
82. Wang, Z.Z., et al., *Comparison of clonogenic assay with premature chromosome condensation assay in prediction of human cell radiosensitivity*. World J Gastroenterol, 2006. 12(16): p. 2601-5.
83. Dresselhaus, M.S., et al., *Raman spectroscopy on isolated single wall carbon nanotubes*. Carbon, 2002. 40(12): p. 2043-2061.
84. Dresselhaus, M.S., et al., *Raman spectroscopy on one isolated carbon nanotube*. Physica B-Condensed Matter, 2002. 323(1-4): p. 15-20.
85. Keszler, A.M., et al., *Characterisation of carbon nanotube materials by Raman spectroscopy and microscopy - A case study of multiwalled and singlewalled samples*. Journal of Optoelectronics and Advanced Materials, 2004. 6(4): p. 1269-1274.
86. Matthäus, C., et al., *Raman and infrared microspectral imaging of mitotic cells*. Applied Spectroscopy, 2006. 60(1): p. 1-8.
87. Raman, C.V. and K.S. Krishnan, *A New Type of Secondary Radiation*. Nature, 1928. 121.
88. Domingo, C. and S. Montero, *Raman Intensities of Sulfur Alpha-S8*. Journal of Chemical Physics, 1981. 74(2): p. 862-872.
89. Dobrosz, P., et al., *The use of Raman spectroscopy to identify strain and strain relaxation in strained Si/SiGe structures*. Surface & Coatings Technology, 2005. 200(5-6): p. 1755-1760.
90. Lyng, F.M., et al., *Vibrational spectroscopy for cervical cancer pathology, from biochemical analysis to diagnostic tool*. Exp Mol Pathol, 2007. 82(2): p. 121-9.
91. Meade, A.D., et al., *Growth substrate induced functional changes elucidated by FTIR and Raman spectroscopy in in-vitro cultured human keratinocytes*. Analytical and Bioanalytical Chemistry, 2007. 387(5): p. 1717-1728.
92. Short, K.W., et al., *Raman spectroscopy detects biochemical changes due to proliferation in mammalian cell cultures*. Biophys J, 2005. 88(6): p. 4274-88.
93. Notingher, I., et al., *Spectroscopic study of human lung epithelial cells (A549) in culture: living cells versus dead cells*. Biopolymers, 2003. 72(4): p. 230-40.
94. Sigurdsson, S., et al., *Detection of skin cancer by classification of Raman spectra*. IEEE Trans Biomed Eng, 2004. 51(10): p. 1784-93.
95. Owen, C.A., et al., *In vitro toxicology evaluation of pharmaceuticals using Raman micro-spectroscopy*. J Cell Biochem, 2006. 99(1): p. 178-86.
96. Scholz, M., et al., *Metabolite fingerprinting: detecting biological features by independent component analysis*. Bioinformatics, 2004. 20(15): p. 2447-2454.
97. Haaland, D.M. and E.V. Thomas, *Partial least-squares methods for spectral analyses. 1. Relation to other quantitative calibration methods and the extraction of qualitative information*. Anal. Chem. ; Vol/Issue: 60:11, 1988: p. Pages: 1193-1202.
98. Beattie, R.J., et al., *Preliminary investigation of the application of Raman spectroscopy to the prediction of the sensory quality of beef silverside*. Meat Science, 2004. 66(4): p. 903-913.

99. Herzog, E., et al., *A new approach to the toxicity testing of carbon-based nanomaterials - The clonogenic assay.* Toxicology Letters, 2007. 174(1-3): p. 49-60.
100. Plumb, J.A., *Cell Sensitivity Assays*, in *Cytotoxic Drug Resistance Mechanisms*. 1999. p. 17-23.
101. Casey, A., et al., *Single walled carbon nanotubes induce indirect cytotoxicity by medium depletion in A549 lung cells.* Toxicology Letters, 2008. 179(2): p. 78-84.
102. Casey, A., et al., *Probing the interaction of single walled carbon nanotubes within cell culture medium as a precursor to toxicity testing.* Carbon, 2007. 45(1): p. 34-40.
103. Casey, A., et al., *Interaction of carbon nanotubes with sugar complexes.* Synthetic Metals, 2005. 153(1-3): p. 357-360.
104. Herzog, E., et al., *SWCNT suppress inflammatory mediator responses in human lung epithelium in vitro.* Toxicology and Applied Pharmacology, 2008.
105. Socrates, G., *Infrared and Raman Characteristic Group Frequencies Tables and Charts.* 3rd ed. 2004: John Wiley & Sons. 366.
106. Stone, N., et al., *Raman Spectroscopy for Early Detection of Laryngeal Malignancy: Preliminary Results.* Laryngoscope, 2000. 110(10): p. 1756-1763.
107. Perna, G., et al., *Raman spectroscopy and atomic force microscopy study of cellular damage in human keratinocytes treated with HgCl2.* Journal of Molecular Structure, 2007. 834: p. 182-187.
108. Yehia, H., et al., *Single-walled carbon nanotube interactions with HeLa cells.* 2007. p. 8.
109. Otto, C., N.M. Sijtsema, and J. Greve, *Confocal Raman microspectroscopy of the activation of single neutrophilic granulocytes.* Eur Biophys J, 1998. 27(6): p. 582-9.
110. Bonnier, F., et al., *Three dimensional collagen gels as a cell culture matrix for the study of live cells by Raman spectroscopy.* The Analyst, 2010.
111. Foster, K.A., et al., *Characterization of the A549 Cell Line as a Type II Pulmonary Epithelial Cell Model for Drug Metabolism.* Experimental Cell Research, 1998. 243(2): p. 359-366.
112. Swain, R.J., et al., *Assessment of cell line models of primary human cells by Raman spectral phenotyping.* Biophys J, 2010. 98(8): p. 1703-11.
113. Lin, W., et al., *In vitro toxicity of silica nanoparticles in human lung cancer cells.* Toxicol Applied Pharmacol, 2006. 217(3): p. 252-9.
114. Boesewetter, D., et al., *Alterations of A549 lung cell gene expression in response to biochemical toxins.* Cell Biology and Toxicology, 2006. 22(2): p. 101-118.
115. Jorio, A., et al., *Structural (n, m) Determination of Isolated Single-Wall Carbon Nanotubes by Resonant Raman Scattering.* Physical Review Letters, 2001. 86(6): p. 1118.
116. Choquette, S.J., et al., *Relative Intensity Correction of Raman Spectrometers: NIST SRMs 2241 Through 2243 for 785 nm, 532 nm, and 488 nm/514.5 nm Excitation.* Applied Spectroscopy, 2007. 61: p. 117-129.
117. Schulze, G., et al., *Investigation of Selected Baseline Removal Techniques as Candidates for Automated Implementation.* Applied Spectroscopy, 2005. 59(5): p. 545-574.

118. Schrof, W., J. Klingler, and D. Horn, *Method and apparatus for Raman correlation spectroscopy*. 1999, BASF Aktiengesellschaft (Ludwigshafen, DE): United States.
119. Smith, J.O. *Mathematics of the Discrete Fourier Transform (DFT)*. 2007; Available from: http://www.w3k.org/books/.
120. Brereton, R.G., *Data Analysis for the laboratory and chemical plant*. Chemometrics 2002, New York: Wiley.
121. Jirasek, A., et al., *Accuracy and precision of manual baseline determination*. Applied Spectroscopy, 2004. 58(12): p. 1488-99.
122. Afseth, N.K., V.H. Segtnan, and J.P. Wold, *Raman spectra of biological samples: A study of preprocessing methods*. Applied Spectroscopy, 2006. 60(12): p. 1358-1367.
123. McCreedy, R.J., *Raman Spectroscopy for chemical Analysis* Chemical Analysis, ed. J.D. Winefordner. Vol. 157. 2000, New York I Chichester i Weinheim I Brisbane I Singapore I Toronto: a John Wiley & Sons, Inc., Publication.
124. Savitsky, A. and M.J.E. Golay, *Smoothing and differentiation of data by simplified least squares procedures*. Analytical Chemistry, 1964. 36: p. 1627-1639.
125. Ferraro, J.R. and K. Nakamoto, *Introductory Raman Spectroscopy*. Academic Press. 1994.
126. Orfanidis, S.J., *Introduction to Signal Processing*. Prentice-Hall Signal Processing Series. 1995: Prentice Hall, Upper Saddle River, NJ 07458.
127. Nijssen, A., et al., *Discriminating basal cell carcinoma from perilesional skin using high wave-number Raman spectroscopy*. J Biomed Optics, 2007. 12(3): p. 034004.
128. Mahadevan-Jansen, A., et al., *Near-infrared Raman spectroscopy for in vitro detection of cervical precancers*. Photochem Photobiol, 1998. 68(1): p. 123-32.
129. Notingher, I., et al., *In situ non-invasive spectral discrimination between bone cell phenotypes used in tissue engineering*. J Cell Biochem, 2004. 92(6): p. 1180-92.
130. Bronstein, I.N. and K.A. Semedyayev, *Handbook of Mathemathics*. 20. ed. 1997: Springer.
131. Wrangsjoe, A. and H. Knutsson, *Histogram Filters for Noise Reduction*, in *Proceedings of the SSAB Symposium on Image Analysis*. 2003.
132. Kaiser, J.F. and W.A. Reed, *Data smoothing using low-pass digital filters*. Review of Scientific Instruments, 1977. 48(11): p. 1447-1457.
133. Bulmer, J.T., et al., *Baseline Modeling with a Computerized Raman System*. Applied Spectroscopy, 1975. 29(6): p. 506-511.
134. Haight, S.M. and D.T. Schwartz, *Automated Image Background Removal in Line-Imaging Raman Spectroscopy*. Applied Spectroscopy, 1997. 51(7): p. 930-938.
135. Heraud, P., et al., *Effects of pre-processing of Raman spectra on in vivo classification of nutrient status of microalgal cells*. Journal of Chemometrics, 2006. 20(5): p. 193-197.
136. Chan, J.W., et al., *Micro-Raman spectroscopy detects individual neoplastic and normal hematopoietic cells*. Biophys J, 2006. 90(2): p. 648-56.

137. Guo, J., et al., *Raman spectroscopic investigation on the interaction of malignanthepatocytes with doxorubicin.* Biophysical Chemistry, 2009. 140(1-3): p. 57-61.
138. Huang, Z.W., et al., *Effect of formalin fixation on the near-infrared Raman spectroscopy of normal and cancerous human bronchial tissues.* International Journal of Oncology, 2003. 23(3): p. 649-655.
139. Pelletier, M.J., *Quantitative analysis using Raman spectrometry.* Applied Spectroscopy, 2003. 57(1): p. 20A-42A.
140. Weisstein, E.W. *Trapezoidal Rule.* 2008; Available from: http://mathworld.wolfram.com/TrapezoidalRule.html.
141. Notingher, I., *Raman Spectroscopy cell-based Biosensors.* Sensors, 2007. 7(8): p. 1343-1358.
142. Young, I.T., J.J. Gerbrands, and L.J. van Vliet, *Fundamentals of Image Processing*, in *Digital Image Processing / Digital Image Analysis*. 1995, Delft University of Technology: Delft.
143. Baker, M.J., et al., *Discrimination of prostate cancer cells and non-malignant cells using secondary ion mass spectrometry.* The Analyst, 2008. 133(2): p. 175-9.
144. Ariai, J. and S.R.P. Smith, *The Raman-Spectrum and Analysis of Phonon Modes in Sodalite.* Journal of Physics C-Solid State Physics, 1981. 14(8): p. 1193-1202.
145. Boydston-White, S., et al., *Cell-cycle-dependent variations in FTIR micro-spectra of single proliferating HeLa cells: Principal component and artificial neural network analysis.* Biochim Biophys Acta, 2006. 1758(7): p. 908-14.
146. Romeo, M., et al., *Infrared micro-spectroscopic studies of epithelial cells.* Biochim Biophys Acta, 2006.
147. Puppels, G.J., et al., *Laser Irradiation and Raman-Spectroscopy of Single Living Cells and Chromosomes - Sample Degradation Occurs with 514.5Nm but Not with 660Nm Laser-Light.* Experimental Cell Research, 1991. 195(2): p. 361-367.
148. Kuzmany, H., et al., *Spectroscopic analysis of different types of single-wall carbon nanotubes.* Europhysics Letters, 1998. 44(4): p. 518-524.
149. Ida, T., M. Ando, and H. Toraya, *Extended pseudo-Voigt function for approximating the Voigt profile.* Journal of Applied Crystallography, 2000. 33(6): p. 1311-1316.
150. Evett, I.W., P.D. Gill, and J.A. Lambert, *Taking account of peak areas when interpreting mixed DNA profiles.* J Forensic Sci, 1998. 43(1): p. 62-9.
151. Ian Marshall, S.D.B.J.H.A.M.J.M.W.K.J.F.J.S., *Choice of spectroscopic lineshape model affects metabolite peak areas and area ratios.* 2000. p. 646-649.
152. Watts, P.J., et al., *Fourier transform-Raman spectroscopy for the qualitative and quantitative characterization of sulfasalazine-containing polymeric microspheres.* Pharm Res, 1991. 8(10): p. 1323-8.
153. Sikorska, E., et al., *Monitoring beer during storage by fluorescence spectroscopy.* Food Chemistry, 2006. 96(4): p. 632-639.
154. Wold, S., H. Martens, and H. Wold, *The Multivariate Calibration-Problem in Chemistry Solved by the Pls Method.* Lecture Notes in Mathematics, 1983. 973: p. 286-293.
155. Martens, H.a.N., T, *Multivariate Calibration.* 1989: Wiley.

156. Leardi, R., *Application of genetic algorithm-PLS for feature selection in spectral data sets.* J Chemometr J Chemometr, 2000. 14(5-6): p. 643-655.
157. Petibois, C., et al., *Analytical performances of FT-IR spectrometry and imaging for concentration measurements within biological fluids, cells, and tissues.* The Analyst, 2006. 131(5): p. 640-647.
158. Zhu, Q., R.G. Quivey, and A.J. Berger, *Raman Spectroscopic Measurement of Relative Concentrations in Mixtures of Oral Bacteria.* Applied Spectroscopy, 2007. 61: p. 1233-1237.
159. Leardi, R., *Application of genetic algorithm-PLS for feature selection in spectral data sets.* Journal of Chemometrics, 2000. 14(5-6): p. 643-655.
160. Leardi, R. and A.L. Gonzalez, *Genetic algorithms applied to feature selection in PLS regression: how and when to use them.* Chemometrics and Intelligent Laboratory Systems, 1998. 41(2): p. 195-207.
161. Jouanrimbaud, D., et al., *Genetic Algorithms as a Tool for Wavelength Selection in Multivariate Calibration.* Analytical Chemistry, 1995. 67(23): p. 4295-4301.
162. Everitt, B.S., *Cluster Analysis.* 1 ed. 1993: John Wiley & Sons Canada, Limited. 170.
163. Wu, J.-d., et al., *Hierarchical Cluster Analysis Applied to Workers Exposures in Fiberglass Insulation Manufacturing.* 1999. p. 43-55.
164. Hyvarinen, A. and E. Oja, *Independent component analysis: algorithms and applications.* Neural Netw, 2000. 13(4-5): p. 411-30.
165. Silveira, L., et al., *Independent Component Analysis Applied to Raman Spectra for Classification of <i>In Vitro</i> Human Coronary Arteries.* 2008, Taylor & Francis. p. 134 - 145.
166. Vrabie, V., et al., *Independent component analysis of Raman spectra: Application on paraffin-embedded skin biopsies.* Biomedical Signal Processing and Control, 2007. 2(1): p. 40-50.
167. Lasch, P., et al., *Infrared spectroscopy of human cells and tissue: detection of disease.* Technol Cancer Res Treat, 2002. 1(1): p. 1-7.
168. van Manen, H.J. and C. Otto, *Hybrid confocal Raman fluorescence microscopy on single cells using semiconductor quantum dots.* Nano Lett, 2007. 7(6): p. 1631-6.
169. Doorn, S., et al., *Raman Spectral Imaging of a Carbon Nanotube Intramolecular Junction.* Physical Review Letters, 2005. 94: p. 16802-4.
170. Knief, P., et al., *Raman spectroscopy--a potential platform for the rapid measurement of carbon nanotube-induced cytotoxicity.* The Analyst, 2009. 134(6): p. 1182-91.
171. Casey, A., et al., *Spectroscopic analysis confirms the interactions between single walled carbon nanotubes and various dyes commonly used to assess cytotoxicity.* Carbon, 2007. 45(7): p. 1425-1432.
172. Hasegawa, T., J. Nishijo, and J. Umemura, *Separation of Raman spectra from fluorescence emission background by principal component analysis.* Chemical Physics Letters, 2000. 317(6): p. 642-646.
173. Frost, K.J. and R.L. McCreery, *Calibration of Raman spectrometer instrument response function with luminescence standards: An update.* Applied Spectroscopy, 1998. 52(12): p. 1614-1618.
174. Bose, S.M., S. Gayen, and S.N. Behera, *Theory of the tangential G-band feature in the Raman spectra of metallic carbon nanotubes.* Physical Review B, 2005. 72(15).

175. Stone, N., et al., *Near-infrared Raman spectroscopy for the classification of epithelial pre-cancers and cancers.* Journal of Raman Spectroscopy, 2002. 33(7): p. 564-573.
176. Puppels, G.J., et al., *Raman Microspectroscopic Approach to the Study of Human Granulocytes.* Biophysical Journal, 1991. 60(5): p. 1046-1056.
177. Notingher, I., et al., *In situ spectral monitoring of mRNA translation in embryonic stem cells during differentiation in vitro.* Anal Chem, 2004. 76(11): p. 3185-93.
178. Lu L., Ward M., and M. A., *Chemical Analysis of Powdered Metallurgical Slags by X-ray Fluorescence Spectrometry* Iron Steel Inst Jpn, 2003. 43(12): p. 1940-1946.
179. Giordani, S., et al., *Debundling of single-walled nanotubes by dilution: observation of large populations of individual nanotubes in amide solvent dispersions.* J Phys Chem B, 2006. 110(32): p. 15708-18.
180. Schulze, C., et al., *Not ready to use - overcoming pitfalls when dispersing nanoparticles in physiological media.* Nanotoxicology, 2008. 2(2): p. 51-61.
181. Verrier, S., et al., *In situ monitoring of cell death using Raman microspectroscopy.* 2004. p. 157-162.
182. Meade, A.D., et al., *Studies of chemical fixation effects in human cell lines using Raman microspectroscopy.* Anal Bioanal Chem, 2010.
183. March, J., *Advanced Organic Chemistry.* 1985, New York: Wiley.
184. Swain, R.J. and M.M. Stevens, *Raman microspectroscopy for non-invasive biochemical analysis of single cells.* Biochem Soc Trans, 2007. 35(Pt 3): p. 544-9.
185. Lee, M.S., et al., *The novel combination of chlorpromazine and pentamidine exerts synergistic antiproliferative effects through dual mitotic action.* Cancer Res, 2007. 67(23): p. 11359-67.
186. Huang, Y., et al., *Serum withdrawal and etoposide induce apoptosis in human lung carcinoma cell line A549 via distinct pathways.* Apoptosis, 1997. 2(2): p. 199-206.
187. Erikstein, B.S., et al., *Cellular stress induced by resazurin leads to autophagy and cell death via production of reactive oxygen species and mitochondrial impairment.* Journal of Cellular Biochemistry, 2010. 111(3): p. 574-584.
188. Baskar Rao, P. and H.J. Byrne, *Investigation of Sodium Dodecyl Benzene Sulfonate Assisted Dispersion and Debundling of Single-Wall Carbon Nanotubes.* The Journal of Physical Chemistry C, 2007. 112(2): p. 332-337.
189. Cheng, Q., et al., *Effect of Solvent Solubility Parameters on the Dispersion of Single-Walled Carbon Nanotubes.* The Journal of Physical Chemistry C, 2008. 112(51): p. 20154-20158.
190. Movasaghi, Z., S. Rehman, and I.U. Rehman, *Raman spectroscopy of biological tissues.* Applied Spectroscopy Reviews, 2007. 42(5): p. 493-541.
191. Notingher, I., et al., *In situ characterisation of living cells by Raman spectroscopy.* Spectroscopy-an International Journal, 2002. 16(2): p. 43-51.
192. Holman, H.Y., et al., *IR spectroscopic characteristics of cell cycle and cell death probed by synchrotron radiation based Fourier transform IR spectromicroscopy.* Biopolymers, 2000. 57(6): p. 329-35.
193. Donaldson, K., et al., *Combustion-derived nanoparticles: a review of their toxicology following inhalation exposure.* Part Fibre Toxicol, 2005. 2: p. 10.

194. Galanzha, E.I., et al., *Skin backreflectance and microvascular system functioning at the action of osmotic agents.* Journal of Physics D: Applied Physics, 2003. 36(14): p. 1739-1746.
195. Lopez-Huertas, E., et al., *Stress induces peroxisome biogenesis genes.* EMBO J, 2000. 19(24): p. 6770-6777.
196. Löffler G., P.P.E., *Biochemie und Pathobiochemie.* 5. ed. Vol. 1. 1975, Berlin Heidelberg New York: Springer Verlag. 1155.
197. Valko, M., et al., *Free radicals and antioxidants in normal physiological functions and human disease.* Int J Biochem Cell Biol, 2007. 39(1): p. 44-84.
198. Rahman, I., et al., *4-Hydroxy-2-nonenal, a specific lipid peroxidation product, is elevated in lungs of patients with chronic obstructive pulmonary disease.* 2002. p. 490 - 495.
199. Rahman, I. and W. MacNee, *Oxidative stress and regulation of glutathione in lung inflammation.* Eur Respir J, 2000. 16(3): p. 534-54.
200. Lanone, S. and J. Boczkowski, *Biomedical applications and potential health risks of nanomaterials: molecular mechanisms.* Curr Mol Med, 2006. 6(6): p. 651-63.
201. Zhang, X., et al., *Effect of pentoxifylline in reducing oxidative stress-induced embryotoxicity.* J Assist Reprod Genet, 2005. 22(11-12): p. 415-7.
202. Halliwell, B., M.V. Clement, and L.H. Long, *Hydrogen peroxide in the human body.* FEBS Lett, 2000. 486(1): p. 10-3.
203. Fenoglio, I., et al., *Reactivity of carbon nanotubes: Free radical generation or scavenging activity?* Free Radical Biology and Medicine, 2006. 40(7): p. 1227-1233.
204. Choi, S.-J., J.-M. Oh, and J.-H. Choy, *Toxicological effects of inorganic nanoparticles on human lung cancer A549 cells.* Journal of Inorganic Biochemistry, 2009. 103(3): p. 463-471.
205. Kim, C.H., et al., *Protein kinase C-ERK1/2 signal pathway switches glucose depletion-induced necrosis to apoptosis by regulating superoxide dismutases and suppressing reactive oxygen species production in A549 lung cancer cells.* Journal of Cellular Physiology, 2007. 211(2): p. 371-385.
206. Lodish, H., *Molecular Cell Biology.* 2008, San Francisco: W.H. Freeman.
207. Chang, W.-T., et al., *Real-time molecular assessment on oxidative injury of single cells using Raman spectroscopy.* Journal of Raman Spectroscopy, 2009. 40(9): p. 1194-1199.
208. Yoshimoto, M., et al., *Phosphatidylcholine vesicle-mediated decomposition of hydrogen peroxide.* Langmuir, 2007. 23(18): p. 9416-22.
209. Valko, M., et al., *Free radicals, metals and antioxidants in oxidative stress-induced cancer.* Chem Biol Interact, 2006. 160(1): p. 1-40.
210. Bayram, H., et al., *Regulation of human lung epithelial cell numbers by diesel exhaust particles.* Eur Respir J, 2006. 27(4): p. 705-13.
211. Ellis, D.I. and R. Goodacre, *Metabolic fingerprinting in disease diagnosis: biomedical applications of infrared and Raman spectroscopy.* The Analyst, 2006. 131(8): p. 875-85.
212. Maquelin, K., et al., *Identification of medically relevant microorganisms by vibrational spectroscopy.* J Microbiol Methods, 2002. 51(3): p. 255-71.
213. Bassan, P., et al., *Resonant Mie scattering in infrared spectroscopy of biological materials--understanding the 'dispersion artefact'.* The Analyst, 2009. 134(8): p. 1586-93.

214. Bassan, P., et al., *Reflection contributions to the dispersion artefact in FTIR spectra of single biological cells.* The Analyst, 2009. 134(6): p. 1171-5.
215. Bassan, P., et al., *Resonant Mie scattering (RMieS) correction of infrared spectra from highly scattering biological samples.* Analyst, 2010. 135(2): p. 268-77.
216. Bonnier, F., et al., *In vitro analysis of immersed human tissues by Raman microspectroscopy.* Journal of Raman Spectroscopy, 2010(in print).
217. Notingher, I., J. Selvakumaran, and L.L. Hench, *New detection system for toxic agents based on continuous spectroscopic monitoring of living cells.* Biosens Bioelectron, 2004. 20(4): p. 780-9.
218. Overton, G. *Spectrometer miniaturization redefines real-time sensing.* PHOTONICS APPLIED 2010 23.08.2010]; Miniaturization of light sources and optical components has shrunk the spectrometer into a portable, handheld form factor that allows the spectrometer to leave the lab and enter the field for real-time material analysis.]. Available from: http://www.optoiq.com/index/photonics-technologies-applications/lfw-display/lfw-article-display/2118518268/articles/optoiq2/photonics-technologies/technology-products/test-measurement/spectrometers-/2010/8/photonics-applied-handheld-spectrometers-spectrometer-miniaturization-redefines-real-time-sensing.html.
219. Gobinet, C., et al., *Preprocessing methods of Raman spectra for source extraction on biomedical samples: application on paraffin-embedded skin biopsies.* IEEE Trans Biomed Eng, 2009. 56(5): p. 1371-82.
220. Luypaert, J., et al., *The effect of preprocessing methods in reducing interfering variability from near-infrared measurements of creams.* Journal of Pharmaceutical and Biomedical Analysis, 2004. 36(3): p. 495-503.
221. Maquelin, K., et al., *Prospective study of the performance of vibrational spectroscopies for rapid identification of bacterial and fungal pathogens recovered from blood cultures.* J Clin Microbiol, 2003. 41(1): p. 324-9.
222. Cooper, S. and M. Gonzalez-Hernandez, *Experimental reconsideration of the utility of serum starvation as a method for synchronizing mammalian cells.* Cell Biol Int, 2009. 33(1): p. 71-7.
223. Wallace, W., et al., *Phospholipid lung surfactant and nanoparticle surface toxicity: Lessons from diesel soots and silicate dusts*, in *Nanotechnology and Occupational Health.* 2007. p. 23-38.
224. Mukherjee, S.P., M. Davoren, and H.J. Byrne, *In vitro mammalian cytotoxicological study of PAMAM dendrimers - towards quantitative structure activity relationships.* Toxicol In Vitro, 2010. 24(1): p. 169-77.
225. Mukherjee, S.P., et al., *Mechanistic studies of in vitro cytotoxicity of poly(amidoamine) dendrimers in mammalian cells.* Toxicol Appl Pharmacol, 2010. 248(3): p. 259-68.
226. Naha, P.C., et al., *Reactive oxygen species (ROS) induced cytokine production and cytotoxicity of PAMAM dendrimers in J774A.1 cells.* Toxicol Appl Pharmacol, 2010. 246(1-2): p. 91-9.
227. Ryman-Rasmussen, J.P., J.E. Riviere, and N.A. Monteiro-Riviere, *Surface coatings determine cytotoxicity and irritation potential of quantum dot nanoparticles in epidermal keratinocytes.* J Invest Dermatol, 2007. 127(1): p. 143-53.

Appendix I

Books

Knief, P. (2010), *"Signal conditioning in Raman Spectroscopy Signals: Is there a way to tackle the baseline problem in Raman Spectroscopy?"*, VDM, ISBN:3639277414

Articles

Meade, A.D., Lyng, F.M., Knief, P., Byrne, H.J., (2007), *"Growth substrate induced functional changes elucidated by FTIR and Raman spectroscopy in in-vitro cultured human keratinocytes"*. Analytical and Bioanalytical Chemistry.

Lyng, F.M., Faolain, E.O., Conroy, J., Meade, A.D., Knief, P., Duffy, B., Hunter, M.B., Byrne, J.M., Keelehan, P., Byrne, H.J., (2007), *"Vibrational spectroscopy for cervical cancer pathology, from biochemical analysis to diagnostic tool"*. Experimental Molecular Pathology,.

Knief, P., Clarke, C., Herzog, E., Davoren, M., Lyng, F.M., Meade, A.D., Byrne, H.J., (2009), *"Raman spectroscopy – a potential platform for the rapid measurement of carbon nanotube-induced cytotoxicity"*, Analyst.

Meade A.D., Clarke, C., Bonnier, F., Poon, K., Garcia, A., Knief, P., Ostrowska, K., Salford, L., Nawaz, H., Lyng F.M., Byrne, H.J., *"Functional and Pathological analysis of biological systems using vibrational spectroscopy with chemometric and heuristic approaches"*, IEEE Conference Proceedings,

Bonnier, F., Meade, A.D., Merza, S., Knief, P., Bhattacharya, K., Lyng, F.M., Byrne, H.J., (2010), *"Three dimensional collagen gels as a cell culture matrix for the study of live cells by Raman spectroscopy"* The Analyst

Bonnier, F., Knief, P., Lim, B., Meade, A. D., Dorney, J., Bhattacharya, K., Lyng, F. M., Byrne, H. J., (2010), *"Imaging live cells grown on a three dimensional collagen matrix using Raman microspectroscopy"*, The Analyst

Bonnier, F., Mehmood, A., Knief, P., Meade, A. D., Hornebeck, W., Lambkin, H., Flynn, K., McDonagh, V., Healy, C., Lee, T.C., Lyng, F. M.,Byrne, H. J., (2010), *"In vitro analysis of immersed human tissues by Raman microspectroscopy"*, Journal of Raman Spectroscopy

Nawaz, H., Byrne, H. J. , Knief, P. , Howe, O., Bonnier, F., Lyng, F.M., Meade, A.D., (2010), *"Evaluation of the potential of Raman microspectroscopy for prediction of chemotherapeutic response to cisplatin in lung adenocarcinoma cells"*, Analyst.

Presentations / Workshops

2007

Oral presentation at 12th ECSBM, Paris, European Conference on the Spectroscopy of Biological Molecules

Poster presentation at Photonics Ireland, Galway, Ireland

Poster presentation at the Joint Conference Nanotechnology, Dublin, Ireland

Poster presentation at the FTIR Worksop RKI Berlin, Germany

Poster presentation at ESF-EMBO Symposium on Probing Inter-actions between Nanoparticles/Biomaterials and Biological Systems, San Feliu de Guixols, Spain

2008

Poster presentation at DASIM Workgroup Meeting, Dublin, Ireland

International Summer School on Synchrotron Infrared Microspectroscopy, Karlsruhe, Germany

Poster presentation at ICPS 2008 Krakow, Poland

Poster presentation at BIGSS Summer School, The Burren, Ireland

Poster presentation at SPEC 2008, Sao Jose Dos Campos, Brazil

2009

Poster presentation at RCSI NBIP Conference, Dublin

Oral presentation at the XLIV Zakopane School of Physics International Symposium, Zakopane, Poland

Poster presentation to 13th ECSBM, Palermo, European Conference on the Spectroscopy of Biological Molecules

Poster presentation at the FTIR Workshop RKI Berlin, Germany

Appendix II

Articles in preparation

Poon, K.W.C., Lyng, F. M., Knief, P., Howe, O., Meade, A. D., Curtin, J.F., Byrne, H.J., Vaughan, J. *"Quantitative reagent-free detection of Fibrinogen Levels by Raman Spectroscopy"*

Knief, P. et al., *"The impact of preprocessing strategies on the explanatory power of multivariate statistics in Raman spectroscopy"*.

Knief, P. et al., *"Applied substrate knowledge - Model based Substrate Removal for Quantitative Analysis in Raman Spectroscopy?"*

Knief, P. et al., *"Probing the toxic effects of ROS on A549 by Raman Spectroscopy"*

Fundamental Matlab code snippets

intensity calibration algorithm

```
function[corrected]=srmcal_pk(signal,srm,factor,wn)
% srmcal_pk
%
% Version 2.1 last modified 06.07.09
%
% Copyright (C) Peter Knief 2009
%
%
% This routine is intended to deliver a luminescence corrected sample
% signal according to the NIST SRM 2243,2242 & 2241 procedures see
% https://srmors.nist.gov/view_cert.cfm?srm=2243 (2241)
%
% Usage : function[corrected]=srmca_pk(signal,srm,factor,wn)
%
% 'signal' is the x,y columnvector of the sample signal 'srm' is the x,y
% columnvector of the srm spectrum at the same settings the
% 'factor' is the multiplier for measurement time of data/srm
% 'wn' is the excitation wavenumber
%
% measurements were conducted x(signal) & x(srm224x) must be the same !
if nargin < 3
    factor=1
end
if nargin < 4
    wn=514;
end
% Coefficients for SRM Correction
% pd(m/p) positive +2sigma or negative -2sigma coefficients
if wn==488
    p=[1.28418E-21 -1.56901E-17 5.75429E-14 -6.11522E-11 6.06655E-08 0 6.9146E-03];
    pd=[3.09685E-22 -4.64692E-18 2.65081E-14 -7.49851E-11 1.20287E-07 -1.08459E-04 0.0490337];
end
if wn==514
    p=[-1.16921E-21 2.38545E-17 -1.70340E-13 4.90077E-10 -4.84706E-07 3.17690E-04 -0.0244612];
    pd=[6.87758E-23 -7.06238E-19 1.93433E-15 -1.49980E-12 1.08168E-08 -3.17886E-05 0.0284858];
end
if wn==532
    p=[9.76795E-18 -9.04836E-14 2.19705E-10 -4.21311E-8 1.22531E-04 0.037014];
    pdp=[9.36292E-18 -8.55445E-14 1.99942E-10 -1.16688E-8 1.0795E-04 0.040358];
    pdm=[1.01730E-17 -9.54227E-14 2.39468E-10 -7.25933E-8 1.37112E-04 0.03367];
end
if wn==785
    p=[1.3535E-01 2.1658E-04 0 1.8936E-10 -9.837E-14 1.2414E-17 0];
    pdp=[1.4221E-01 2.2349E-04 0 1.9434E-10 -1.0331E-13 1.3532E-17 0];
    pdm=[1.2916E-01 2.1016E-04 0 1.8034E-10 -9.099E-14 1.0948E-17 0];
end
if length(signal(:,1))~=length(srm(:,1))
    errordlg('SRM signal does not match the sample spectrum, exiting ')
    Break
end
I_srm=polyval(p,signal(:,1));
S_srm=(srm(:,2))*factor;
I_corr=I_srm./S_srm;
I_srm=polyval(p,signal(:,1));
S_srm=(srm(:,2))*factor;
I_corr=I_srm./S_srm;
[x,ys]=size(signal);
for i=2:ys
    signal(:,i)=signal(:,i)./I_corr;
end
corrected=signal;
```

linearization of spectra

```
function[newosig]=linearize_pk(isignal,precision,plotoff)
%
% linearize_pk
%
% File Version 2.4 02.07.2010
%
% Copyright (C) Peter Knief 2010
%
% Function to linearize spectral data by linear interpolation between the
% measured datapoints and reduction to full decimals dependent on the targeted precision.
%
% usage : function[newosig]=linearize_pk(isignal,precision,plotoff)
%
% plotoff = 1 no visual output
% precision = 1 if not given otherwise
% isignal=[wavenumber spectrum_1 ... spectrum_n] as columnvector
% newosign is returned the same way
%
newsig=zeros(1,2);
if nargin < 2
    precision = 1;
end
if nargin < 3
    plotoff=1;
end
s=0;
%yn = new y value
% p = new x position
%ya = last y value
%yb = next y value
%xb = next x value
%xa = last x value
for j=2:size(isignal,2); % inputsignal more than one
    signal=[isignal(:,1) isignal(:,j)];
xn=ceil(signal(1,1)):precision:floor(signal((size(signal,1)),1));
% xn number of nodes after linearization
for i=1:size(xn,2)
p=xn(i);
    xa=(find(signal(:,1) > p,1)-1); % find xa of signal from the second segment on
    if xa == 0
        xa=signal(1,1);
    end
    xb=(find(signal(:,1) > p,1));    % find xb of signal
    if xb == 0
        xb=signal((size(signal,1)),1);
    end
ya=signal(xa,2);
yb=signal(xb,2);
xa=(signal(xa,1));
xb=(signal(xb,1));
x=p-xa;
y0=ya;
a=(yb-ya)/(xb-xa);
yn=a*x+y0;
newsig(i,1:2)=[p yn];
end
% set last value
newsig(size(newsig,1),2)=signal(size(signal,1),2);
if plotoff == 0
    fig1=figure('name','linearistion Result');
    hold on;
    plot (newsig(:,1),newsig(:,2),'r');
    plot (signal(:,1),signal(:,2));
    hold off;
end
% assining the linearized spectra
```

```
newosig(:,1)=newsig(:,1);
newosig(:,j)=newsig(:,2);
end
```

dynamics assisted moving average noise filtering

```
function[osignal,SNR]=denoise_pk(isignal,plotoff)
%
% Version 1.3 last modified 19.02.2010
%
% Copyright (C) Peter Knief 2010
%
% This denoise filter is basically a weighted median filter over 3 elements
% used to denoise a signal, by measuring the noiseband of a signal
% reducing the maximum signal by half of the noise band value, increasing the minimum
% band by half, leaving peaks out that are over or under the band
% and where the band is broarder than the mean of the band and therefore
% employing the signals dynamics as refinement for true signal
%
% Usage : [osignal,SNR]=denoise_pk(isignal,plotoff)
%
% isignal is the matrix of spectra to denoise
% the first column can be the abscisse if so the program will find it !
%
% plotoff = 1 switches plotting of spectra off (default =0)
%
if nargout < 1
plotoff = 0;
end
if nargin < 2
    plotoff=1;
end
nr_spectra=size(isignal,2);
if continuous_pk(isignal(:,1)) >= 1
    wavenumber=isignal(:,1)';
    nr_spectra=nr_spectra-1;
    k=1;
    l=1;
end
if continuous_pk(isignal(:,1)) == 0
    wavenumber=1:size(isignal(:,1),1);
    nr_spectra=nr_spectra-1;
    k=0;
    l=1;
end
for j=k:nr_spectra
signal=[wavenumber' isignal(:,j+1)];
% lsignal=signal;
maxima=maxima_pk(signal(:,2));
minima=minima_pk(signal(:,2));
minimasig=(interfill_pk(signal(:,2)',minima,1))';
maximasig=(interfill_pk(signal(:,2)',maxima,1))';
maximasig=[signal(:,1) maximasig(:,2)];
minimasig=[signal(:,1) minimasig(:,2)];
noise=(maximasig(:,2)-minimasig(:,2));
noisemean=mean(noise,1);
noise(minima)=noise(minima)*-1;
for i=1:size(noise,1)
    if abs(noise(i))>noisemean
        noise(i)=0;
    end
    if signal(i,2) > maximasig(i,2)
        noise(i)=0;
    end
    if signal(i,2) < minimasig(i,2)
        noise(i)=0;
    end
end
```

```
signal(minima,2)=signal(minima,2)-noise(minima)/2;
signal(maxima,2)=signal(maxima,2)-noise(maxima)/2;
%signal(:,2)=signal(:,2)-min(signal(:,2));
SNR=abs(10*log10(noisemean/max(signal(:,2))));
if plotoff == 0
    fig1=figure('name','denoise Result');
    hold on;
    plot (wavenumber,isignal(:,j+1),'r');
    plot (wavenumber,signal(:,2),'g');
    legend('original','filterd');
    if nr_spectra > 3
        pause
        close all;
    end
end
osignal(:,1)=wavenumber;
osignal(:,j+1)=signal(:,2);
end
end
function[steady]=continuous_pk(data)
[x,y]=size(data);
if y > x
    data=data';
end
[x,y]=size(data);
if y >= 2
    f = errordlg('Input vector was not single', 'Input error', 'modal');
    return
end
stf=0;
stc=0;
cstf=0;
cstc=0;
for i=2:size(data(:,1),1)
    if data(i-1,1) >= data(i,1)
        stf=stf+1;
    end
    if data(i-1,1) <= data(i,1)
        stc=stc+1;
    end
    if data(i-1,1) > data(i,1)
        cstf=cstf+1;
    end
    if data(i-1,1) < data(i,1)
        cstc=cstc+1;
    end
end
steady=0;
if stf+1 == i
    steady=1;
end
if stc+1 == i
    steady=2;
end
if cstf+1 == i
    steady=3;
end
if cstc+1 == i
    steady=4;
end
end
function[res]=interfill_pk(signal,nodes,plotoff)
%
% interfill_pk
%
% sequence to bild a signal based on nodes of the original signal eg to
% create a baseline
%
% file Version 1.0
```

```
%
% usage : function[res]=interfill_pk(signal,nodes,plotoff)
%
% plotoff = 1 no visual output
% signal  = is the y data only
% nodes   = matrix of locations in the signal
% res     = signal based on nodes of the original signal
%
if nargin < 3
    plotoff=0;
end
xystart=[1 signal(1)];res=xystart;
for i=1:length(nodes)
        xi=[xystart(1,1)+1:1:(nodes(i)-1)];
        y=[xystart(1,2);signal(nodes(i))];
        x=[xystart(1,1);nodes(i)];
        yi=interp1(x,y,xi);
        out=[xi' yi'];
        xystart=[nodes(i) signal(nodes(i))];
        res=[res;out;xystart];
end;
        xleft=[nodes(i)+1:1:length(signal)];
        sigleft=[xleft;signal(xleft)];
        res=res';
        res=[res sigleft];
        clear 'yi';clear 'y';clear 'xystart';clear 'xleft';clear 'xi';clear 'x';clear 'sigleft';clear 'out';clear 'i';
if plotoff==0
        plot(res(1,:),res(2,:));
end
end
function[minima]=minima_pk(signal)
%find Minima value left & right are bigger than value
j=1;
for i=2:(length(signal)-1)
    c(i,1)=i ;
    if signal(i-1) > signal(i) && signal(i) < signal(i+1)
        minima(j)=i;
        j=j+1;
    end
end
end
function[maxima]=maxima_pk(signal)
maxima=[];
%find Maxima value left & right are bigger than value
j=1;
for i=2:(length(signal)-1)
    c(i,1)=i ;
    if signal(i-1) < signal(i) && signal(i) > signal(i+1)
        maxima(j)=i;
        j=j+1;
    end
end
end
```

histogram assisted noise removal

```
function[output]=hdenoise_pk(isignal,bin,plotoff)
%
% hdenoise_pk
%
% sequence to remove noise by asessing the historamm of areas underneth
% the peaks
%
% file Version 1.4 last modified 04.05.2010
%
% Copyright (C) Peter Knief 2009
%
```

```
% usage : function[output]=hdenoise_pk(signal,bin,plotoff)
%
% plotoff = 1 no visual output default = 1
% signal  = is the [x y0 y1 ... ] (spectral) data
% bin     = startparameter to bin in histogramm to use for filtering
% (default = 2)
% output  = signal after noisereduction
%
%
if nargin < 3
    plotoff = 1;
end
if nargin < 2
    bin = 2;
end
output(:,1)=isignal(:,1);
for w=2:size(isignal,2)
    signal=[isignal(:,1) isignal(:,w)];
min_p=(minima_pk(signal(:,2)));
% assuming a peak is between two local minima !
background=interfill_pk(signal(:,2)',min_p,1);
spikysignal= signal(:,2)-background(2,:)';
locations=(minima_pk(spikysignal))';% find(spikysignal==0);
npeaks=size(locations,1)-1;
for i=1:(npeaks) % Area of the peaks
    peak_a(i)=abs(sum(spikysignal(locations(i,1):locations(i+1,1),1))); % Area of the peaks
end
if bin >= npeaks
    bin = npeaks;
end
 s_peak_a=sort(peak_a);
 [n,xout]=hist(peak_a,npeaks);
 bins=sum(n(1:bin));
 for i=1:npeaks % for each peak
    if peak_a(i) <= s_peak_a(bins)
        sps_mean(i)=peak_a(i)/(locations(i+1,1)-locations(i,1));
        spikysignal(locations(i,1):locations(i+1,1),1)= sps_mean(i) ; %mean(spikysignal(locations(i,1):locations(i+1,1),1));
    end
 end
output(:,w)=(background(2,:)'+spikysignal);
if plotoff==0
    h=figure();
    subplot(211);hold on;hist(peak_a,npeaks);
    xlabel('peak area [a.u.]');ylabel('occurrence [counts]');
    axis('tight');
    [xl]=axis;
    nb=size(xout,2);
    plot([(xl(1,2)/nb)*bin (xl(1,2)/nb)*bin],[xl(1,3) xl(1,4)],'r');
    legend('original','bin level');
    subplot(212);plot(signal(:,1),[signal(:,2) output(:,w)]);
    xlabel('Wavenumbers [cm^-^1]');ylabel('Intensity [a.u.]');
    legend('original','filtered');
%   background(1,:)=((background(1,:))'+min(signal(:,1)))';
%   hold on;plot(background(1,:),background(2,:),'g');
%   % [SNR(w-1,:),nbin]
end
End
end
function[res]=interfill_pk(signal,nodes,plotoff)
%
% interfill_pk
%
% sequence to bild a signal based on nodes of the original signal eg to
% create a baseline
%
% file Version 1.1
%
% Copyright (C) Peter Knief 2010
%
```

```
% usage : function[res]=interfill_pk(signal,nodes,plotoff)
%
% plotoff = 1 no visual output
% signal  = is the y data only
% nodes   = matrix of locations in the signal
% res     = signal based on nodes of the original signal
%
if nargin < 3
    plotoff=0;
end
if nodes(1) == 1;
    nodes=nodes(2:end); % startvalue is 1 anyway !
end
xystart=[1 signal(1)];res=xystart;
for i=1:length(nodes)
        xi=[xystart(1,1)+1:1:(nodes(i)-1)];
        y=[xystart(1,2);signal(nodes(i))];
        x=[xystart(1,1);nodes(i)];
        yi=interp1(x,y,xi);
        out=[xi' yi'];
        xystart=[nodes(i) signal(nodes(i))];
        res=[res;out;xystart];
end;
        xleft=[nodes(i)+1:1:length(signal)];
        sigleft=[xleft;signal(xleft)];
        res=res';
        res=[res sigleft];
        clear 'yi';clear 'y';clear 'xystart';clear 'xleft';clear 'xi';clear 'x';clear 'sigleft';clear 'out';clear 'i';
if plotoff==0

        plot(res(1,:),res(2,:));
end
end
function[minima]=minima_pk(signal)
%find Minima value left & right are bigger than value
j=1;
maxnode = max(size(signal));
for i=2:(length(signal)-1)
    c(i,1)=i ;
    if signal(i-1) > signal(i) && signal(i) < signal(i+1)
        minima(j)=i;
        j=j+1;
    end
    if signal(i) > signal(i-1) && i == 2
        minima(j)=(i-1);
        j=j+1;
    end
    if signal(i) > signal(i+1) && (i+1) == maxnode
        minima(j)=(i+1);
        j=j+1;
    end
end
end
```

rubberband baseline removal

```
function[nsignal,rubber,nodes]=rubberband_pk(isignal,precision,plotoff,debug)

% rubberband_pk
%
% Version 1.7 last modified 06.05.2010
%
% Copyright (C) Peter Knief 2010
%
% This function is intended to remove a baseline in a rubberband fashion
%
% that isstretched over the nodes representet by the minima
```

```
%
% it works only if the signal is to some extent v-shaped
%
% Usage : [osignal,rubber,nodes]=rubberband_pk(signal,precision,plotoff,debug)
%
% isignal   = the is the signal to be treated [x y]
%
% precision = the precision the fit default = .01
%
% nsignal   = the new signal after baseline removal
%
% rubber    = the removed baseline
%
% nodes     = the locations where the rubberband adheres to the signal
%

if nargin < 4
debug=0;
end

if nargin <3
    plotoff=1;
end

if nargin <2
    precision=.01;
end

clear rubber osignal nodes nnodes;

if size(isignal,1) > size(isignal,2)

    isignal=isignal';

end

signal=isignal(2,:)';

if max(match_pk(signal,min(signal))) > 1 && max(match_pk(signal,min(signal))) < length (signal)
% minimum is somewhere in the middle

        if debug == 1
        'in the middel'
        end

          cnodes=match_pk(signal,min(signal));
          lsignal=signal(1:min(cnodes),:);
          [lnodes]=lrubberband_pk(lsignal,precision,debug);
          lnodes=lnodes(1,2:size(lnodes,2));
          rsignal=flipdim(signal(max(cnodes):size(signal,1)),1);
          [rnodes]=lrubberband_pk(rsignal,precision,debug);
          rnodes=size(rsignal,1)-rnodes+max(cnodes);
          % the right nodes are flipped too
          % rnodes=rnodes(1,1:(size(rnodes,2)-1))

          nnodes=[lnodes cnodes' rnodes];
          nnodes=delduplicates(nnodes);
          nnodes=sort(nnodes);


end
if max(match_pk(signal,min(signal))) == 1
% minimum is left
    if debug==1
       'minimum is left'
    end

    rsignal=flipdim(signal,1);
    [nnodes]=lrubberband_pk(rsignal,precision,debug);
```

```
    nnodes=length(signal)-nnodes+1;
    nnodes=delduplicates(nnodes);
end

if max(match_pk(signal,min(signal))) == length (signal)
% minimum is right
    if debug == 1
        'minimum right'
    end

    [nnodes]=lrubberband_pk(signal,precision,debug);
    nnodes=sort(nnodes);
    nnodes=delduplicates(nnodes);
end

nnodes=nnodes(2:(length(nnodes)));

try
    rubber=interfill_pk(signal',nnodes);
catch
    precision=precision/10;
    [nsignal,rubber,nnodes]=rubberband_pk(isignal,precision,plotoff);
end

nsignal=(signal'-rubber(2,:));

if abs(min(nsignal)) > (1e-12)
% precision was to little
    if debug==1
        'precision too little'
    end
    precision=precision/10;
    [nsignal,rubber,nnodes]=rubberband_pk(isignal,precision,plotoff);
end

nsignal=nsignal(size(nsignal,1),:);

nsignal=nsignal-(min(nsignal));

signal=signal';

if plotoff==0
    plot(signal,'r');
    hold on;
    plot(rubber(2,:)');
end


if size(rubber,1) > size(rubber,2)

    rubber=rubber';

end

rubber(1,:)=isignal(1,:);
rubber=rubber';
nodes=nnodes';

end


function[lnode]=lrubberband_pk(signal,precision,debug)

if debug == 1
'lrubberband'
end
if nargin<2
    precision=.01;
end
```

```
% remove y offset
signal=signal-min(signal);

%finding startpoint

xs= max(match_pk(signal,min(signal)));

% defining left part
xl=1;


%% stretch rubberband left
% a=slope (left side negative !
a=0;
% i=counter
i=1;
lnode(1)=xs;
% yl=signal amplitude
% xm= pisition of minimum
% xn=position of new node
xm=xs;
xn=xm;
% difference = signal-linefit

while xm~= xl

    % as long as no new node (xn) is found
    while xn == xm
    % increase slope
    a=a+precision;

    yl=(-a*[xl:xm])';
    difference=signal(xl:xm)-yl;
    xn=match_pk(difference,min(difference));

    end

 i=i+1;
 a=0;
 lnode(i)=xn(1);

 if debug==1
 lnode(i)
 end
 xm=xn;
end
end

function[bestmatch,error]=match_pk(source,lookfor)
tmp=source-lookfor;
bestmatch=find (abs(tmp)==min(abs(tmp)));
error=source(bestmatch)-lookfor;
end

function[res]=interfill_pk(signal,result)
xystart=[1 signal(1)];res=xystart;
for i=1:length(result)
      xi=[xystart(1,1)+1:1:(result(i)-1)];
      y=[xystart(1,2);signal(result(i))];
      x=[xystart(1,1);result(i)];
      yi=interp1(x,y,xi);
      out=[xi' yi'];
      xystart=[result(i) signal(result(i))];
      res=[res;out;xystart];
end;
      xleft=[result(i)+1:1:length(signal)];
      sigleft=[xleft;signal(xleft)];
      res=res';
      res=[res sigleft];
```

```
        clear 'yi';clear 'y';clear 'xystart';clear 'xleft';clear 'xi';clear 'x';clear 'sigleft';clear 'out';clear 'i';
end
function[data]=delduplicates(inpdata)
j=1;
data(1,j)= inpdata(1,1);

for i=2:size(inpdata,2)

    if inpdata(1,i-1) ~= inpdata(1,i)
        j=j+1;
        data(1,j)=inpdata(1,i);
    end
end
end
```

automated simultaneous curve fitting with multiple mixed G-L curves

```
function[output,par]=msglfit_pk(signals,peaks,preknown,plotoff,sf)
% msglfit_pk
%
% Function to fit multiple Gaussian &  a Lorenzian peaks to a set of signals
% version 4.1 Copyright by Peter Knief last modified 06.06.2010 now
% optional including a linear + a exponential + offset function
%
% USAGE [output,par]=msglfit_pk(signals,peaks,preknown,plotoff,sf)
%
% signals is columnvector [x y1 y2 ...]
% peaks is the peaks to fit to
% preknown is a vector of precnown facts [up ua uc ug lp la lc lg] margins
% plotoff plotting of the results
% sf toggles supportfunction
%
% output = [p1 a1 c1 g1 .... pn an cn gn q r s u v]
%
% pn = fitted peak position
% an = amplitude parameter
% cn = width of peak
% gn = weight between G/L
%
% q  = slope of linear function
% r  = multiplyer of exponential function
% s  = exponential fraction numerator coefficient
% u  = exponential fraction denominator
% v  = offset
%
%
% par = all fit parameters in structure
%
if nargin < 5
   sf=0 ;
end
if nargin < 4
   plotoff = 0;
end
if nargin < 3
   preknown = 5;
end
for k = 2: size(signals,2)
   j=(k-1);
   signal=[[signals(:,1) signals(:,k)]];
result=[];
   [cfun,gof,my]=sglfit_pk(signal,peaks,preknown,plotoff,sf);
   par(j).gof=gof;
   par(j).cfun=cfun;
% collecting the fit parameters
for i=1:size(peaks,2)
   a=cfun.(strcat('a',num2str(i)));
   c=cfun.(strcat('c',num2str(i)));
```

```
            g=cfun.(strcat('g',num2str(i)));
            p=cfun.(strcat('p',num2str(i)));
            result=[result p a c g];
        end
        % assing the support function paramaters
        if sf ==1
        result=[result cfun.q cfun.r cfun.s cfun.u cfun.v];
        end
        output(j,:)=result;
        if sf==0 && plotoff == 0
            glfunc(signal(:,1),result,plotoff);
        end
        if sf==1 && plotoff == 0
            glfunc(signal(:,1),result(1,1:end-5),plotoff);
        end
    end
end
function[cfun,gof,my]=sglfit_pk(signal,peaks,preknown,plotoff,sf)
t0=clock;
x=signal(:,1);
my=min(signal(:,2));
y=signal(:,2);%-my;
if size(preknown,2)== 8 && size(preknown,1) == 1
    % parameter valid for all peaks
    for i=1:size(peaks,2) %size(peaks,2)==1
        ul(1,i)=preknown(i,2);% upper limit amplitude
        ul(2,i)=preknown(i,3);% upper limit width
        ul(3,i)=preknown(i,4);% upper limit weighting g/l
        ul(4,i)=preknown(i,1);% upper limit position
        ll(1,i)=preknown(i,6);% lower limit amplitude
        ll(2,i)=preknown(i,7);% lower limit width
        ll(3,i)=preknown(i,8);% lower limit weighting g/l
        ll(4,i)=preknown(i,5);% lower limit position
        sp(1,i)=y(match_pk(x,peaks(i)));% amplitude starting point
        sp(2,i)=mean([preknown(i,7) preknown(i,3)]);% starting point width
        sp(3,i)=mean([preknown(i,8) preknown(i,4)]);% starting point weighting g/l
        sp(4,i)=peaks(i);% starting point position
%       ll(1,i)=0;% lower limit amplitude
%       ll(2,i)=0;% lower limit width
%       ll(3,i)=0;% lower limit weighting g/l
%       ll(4,i)=peaks(i)-precision;% lower limit position
%       sp(1,i)=y(match_pk(x,peaks(i)));% amplitude starting point
%       sp(2,i)=50;% starting point width
%       sp(3,i)=.5;% starting point weighting g/l
%       sp(4,i)=peaks(i);% starting point position
%       ul(1,i)=inf;% upper limit amplitude
%       ul(2,i)=inf;% upper limit width
%       ul(3,i)=1;% upper limit weighting g/l
%       ul(4,i)=peaks(i)+precision;% upper limit position
    end
    multiplier=500;% limited knowledge about the curve
end
if size(preknown,2)== 1 && size(preknown,1) == 1
    % parameter is just precision
    for i=1:size(peaks,2)
        ll(1,i)=0;% lower limit amplitude
        ll(2,i)=0;% lower limit width
        ll(3,i)=0;% lower limit weighting g/l
        ll(4,i)=peaks(i)-preknown;% lower limit position
        sp(1,i)=y(match_pk(x,peaks(i)));% amplitude starting point
        sp(2,i)=50;% starting point width
        sp(3,i)=.5;% starting point weighting g/l
        sp(4,i)=peaks(i);% starting point position
        ul(1,i)=inf;% upper limit amplitude
        ul(2,i)=inf;% upper limit width
        ul(3,i)=1;% upper limit weighting g/l
        ul(4,i)=peaks(i)+preknown;% upper limit position
    end
    multiplier=1000;% totaly no knowledge about the curve
```

```
    end
    if size(preknown,1) == size(peaks,2)% && size(preknown,1) ~= size(preknown,2)
        % diffrent parameter for every peak
        % display('good knowledge');
        for i=1:size(peaks,2)
            %preknown (up ua uc ug lp la lc lg)
            ul(1,i)=preknown(i,2);% upper limit amplitude
            ul(2,i)=preknown(i,3);% upper limit width
            ul(3,i)=preknown(i,4);% upper limit weighting g/l
            ul(4,i)=preknown(i,1);% upper limit position
            ll(1,i)=preknown(i,6);% lower limit amplitude
            ll(2,i)=preknown(i,7);% lower limit width
            ll(3,i)=preknown(i,8);% lower limit weighting g/l
            ll(4,i)=preknown(i,5);% lower limit position
            sp(1,i)=y(match_pk(x,peaks(i)));% amplitude starting point
            sp(2,i)=mean([preknown(i,7) preknown(i,3)]);% starting point width
            sp(3,i)=mean([preknown(i,8) preknown(i,4)]);% starting point weighting g/l
            sp(4,i)=peaks(i);% starting point position
        end
        multiplier=250;% good knowledge about the curve
    end
%% fitting the curve
sglfit=[];
for i=1:size(peaks,2)
j=num2str(i);
sglfit=strcat(sglfit,'+((1-g',j,')*(a',j,'/(1+(x-p',j,')^2/c',j,'^2)+(g',j,'*(a',j,'*exp(-((x-p',j,')/c',j,')^2))))');
end
% adding support fit
if sf==1
sglfit=strcat(sglfit,'+(q*x)+ r*exp((x-s)/u)+ v');
end
ffun = fittype(sglfit);
foptions = fitoptions(ffun);
%
foptions.Algorithm='Trust-Region';
foptions.Robust='LAR';
% parameter        a       c       g       p
foptions.Lower=    [ll(1,:) ll(2,:) ll(3,:) ll(4,:)];
foptions.StartPoint=[sp(1,:) sp(2,:) sp(3,:) sp(4,:)];
tions.Upper=       [ul(1,:) ul(2,:) ul(3,:) ul(4,:)];
%extend f.options by support f.options:
%parameter          a c g p              q  r  s    u     v
if sf==1
foptions.Lower=    [foptions.Lower    -inf -inf 1     0    -inf];
foptions.StartPoint=[foptions.StartPoint 0   0   0     2    0  ];
foptions.Upper=    [foptions.Upper    +inf +inf 2*max(x) 500 +inf];
end
foptions.MaxFunEvals=size(peaks,2)/2*multiplier;
foptions.MaxIter=foptions.MaxFunEvals/2;
foptions.Display='Off';

ffun = fittype(sglfit,'options',foptions);
[cfun,gof] = fit(x,y,ffun);
% etime(clock,t0)
if plotoff == 0
h=figure;
plot(x,[y cfun(x)]);hold on
legend('signal','fit');
bt1=strcat('sse   =',num2str(gof.sse));
bt2=strcat('R^2   =',num2str(gof.rsquare));
bt3=strcat('dfe   =',num2str(gof.dfe));
bt4=strcat('aR^2  =',num2str(gof.adjrsquare));
bt5=strcat('rmse  =',num2str(gof.rmse));
annotation('textbox',[0.157 0.7962 0.09804 0.09613],'String',{bt1,bt2,bt3,bt4,bt5},...
    'FontName','Fixed Miriam Transparent',...
    'FitBoxToText','on');
end
end
function[bestmatch,error]=match_pk(source,lookfori)
```

```
for i=1:size(lookfori,2)
tmp=source-lookfori(i);
bestmatch(i)=find (abs(tmp)==min(abs(tmp)));
error(i)=source(bestmatch(i))-lookfori(i);
end
en
```

I want morebooks!

Buy your books fast and straightforward online - at one of world's fastest growing online book stores! Environmentally sound due to Print-on-Demand technologies.

Buy your books online at
www.morebooks.shop

Kaufen Sie Ihre Bücher schnell und unkompliziert online – auf einer der am schnellsten wachsenden Buchhandelsplattformen weltweit! Dank Print-On-Demand umwelt- und ressourcenschonend produziert.

Bücher schneller online kaufen
www.morebooks.shop

KS OmniScriptum Publishing
Brivibas gatve 197
LV-1039 Riga, Latvia
Telefax +371 686 204 55

info@omniscriptum.com
www.omniscriptum.com

Printed by Books on Demand GmbH, Norderstedt / Germany